ENGINEERING FOR CATS

ENGINEERING FOR CATS

Better the Life of Your Pet
with 10 Cat-Approved Projects

Written and Illustrated by

Mac Delaney

Workman Publishing • New York

For my wife, Isabel

Copyright © 2018 by Mac Delaney

Library of Congress Cataloging-in-Publication Data is available.

ISBN 978-0-7611-8990-9

Workman books are available at special discounts when purchased in bulk for premiums and sales promotions as well as for fund-raising or educational use. Special editions or book excerpts can also be created to specification. For details, contact the Special Sales Director at the address below, or send an email to specialmarkets@workman.com.

Workman Publishing Company, Inc.
225 Varick Street
New York, NY 10014
workman.com

WORKMAN is a registered trademark of Workman Publishing Co., Inc.

Printed in China
First printing July 2018

10 9 8 7 6 5 4 3 2 1

Contents

Prologue

Scene: Conference room in aerospace engineering facility

Manager: Thanks for meeting with me. I need to speak with you about a contract we recently won.

Engineer: Oh, I wasn't aware any bids were recently placed. Is this another single-engine fighter?

Manager: No, not really.

Engineer: Twin-engine?

Manager: No.

Engineer: Quintuple-engine?

Manager: No. Wait, is that a thing?

Engineer: Do you want it to be? Here, let me pull up some sketches.

Manager: No, please listen; we've brought you in because of the great work you and your team did on the X-25 Desert Reagan program.

Engineer: You mean Jetty McJetface?

Manager: No, that's not what it's called.

Engineer: Not according to the voters of the naming contest, sir.

Manager: Moving on. We'd like to discuss this new project with you, but it's going to be a little different from what you've developed in the past. [*pause*] We have actually been funded to take on the long-overdue project of developing a better mousetrap.

Engineer: A septuple-engine fighter that exterminates mice? No problem—what's our schedule look like?

Manager: No, this is not going to be an airplane, just a mousetrap.

Engineer: But you said the requirements were open, so it's not necessarily *not* an airplane?

Manager: Do you think you could build a plane that is cost-effective in exterminating mice?

Engineer: I suppose it depends on the extent of the infestation. How many mice does each product have to handle? 10,000? 100,000?

Manager: I think the trap would have to take care of the mice without blowing up the buildings that they're in.

Engineer: These requirements are less open than you let on.

Manager: Look, is this something we can handle, or are we going to need to bring in some consultants?

Engineer: No, we can handle it. We just need to hammer out the parameters. Let's look at how the current mousetrap can be improved.

Manager: Okay.

Engineer: I think the biggest problem with the standard mousetrap design is that it's stationary. It requires bait like cheese to get the mouse to come to the trap, which is completely ineffective on mice that are lactose intolerant.

Manager: Good point, and if I'm not mistaken, many people just want to get rid of mice because their cheese is being eaten. If people have to use their cheese to set the traps, then it doesn't really solve the problem, does it?

Engineer: Exactly. So we need a trap that does not operate by wasting cheese. Unfortunately I'm not aware of any way to get mice to come to a trap other than cheese, and I don't intend on performing any research to check if there is. So if we can't get the mice to come to the trap, the trap will have to go to the mice.

Manager: I'm with you so far.

Engineer: Imagine a Roomba, but instead of cleaning floors, it hunts mice.

Manager: Okay, but I'm having trouble imagining a Roomba that's quick enough to catch a mouse.

Engineer: Then you're not imagining it correctly. This Roomba would have between four and eight powerful legs to make it fast and agile, built-in weapon systems, and artificial intelligence that allows it to learn how to most effectively terminate the local mouse population.

Manager: That sounds a little dangerous. . . . Could we develop something like that safely?

Engineer: Well, what usually makes most killer robots so dangerous is that they are hardwired to keep us safe.

Manager: Go on.

Engineer: Things in which humans find joy are usually also dangerous, and we're also pretty fragile. So any robot capable of learning eventually figures out that the only way to keep us safe is to enslave all of humanity.

Manager: So . . . we need to make sure our robot also prioritizes our happiness?

Engineer: That could work, but I think it would actually be simpler just to have the robot not care about us at all.

Manager: So what keeps it from killing us accidentally?

Engineer: Well, I think we might be able to make it not powerful enough to kill us.

Manager: Hmm, I suppose I hadn't really considered that.

Engineer: I agree it's a strange concept. If we make it powerful enough to hunt mice, but not powerful enough to kill people, it just might not enslave all of humanity. At least not completely.

Manager: Seems worth pursuing. But if the only things stopping it from killing us are it not caring about us and it being slightly underpowered, won't people be a little hesitant to buy it?

Engineer: That's where aesthetics come in. If we make it stylish enough, then people will overlook any fatal functional flaws. It's called the Apple effect.

Manager: That is a stylish fruit, but not without its flaws, which I'm sure is what you were referring to. . . . I see just one problem in your plan.

Engineer: What's that?

Manager: You want to design a mouse hunter that's mobile, powerful enough to kill a mouse but not powerful enough to kill us, that keeps us safe mostly by ignoring us, and is aesthetically pleasing to people?

Engineer: Yes.

Manager: That's a cat.

Engineer: I see. . . .

Manager: I don't think we can make a robot that competes with a cat.

Engineer: Hmmm, okay. Change of plans. How about we make our mousetrap an actual cat?

Manager: How would we stay in business selling cats?

Engineer: The cats aren't where we make the money, it's the add-ons.

Manager: Like cat refills?

Engineer: Sort of. Look, some people don't even like cats, and a lot of people who *do* aren't taking their cats to their full potential.

Manager: Ah, I see; so we develop steroids for the cats, then.

Engineer: That would work but might conflict with our requirement of making sure the product is not powerful enough to kill humans.

Manager: So we give the humans steroids too? Cancel things out?

Engineer: Let's come back to that. For now, I think the most effective path would be to develop a series of cat add-on products that will improve the human-cat relationship, making the cats happier and more effective mouse-killing machines.

Manager: Perfect, it looks like your team is ready to start . . . [*pauses and dramatically turns head*]

ENGINEERING FOR CATS

(A MORE USEFUL) INTRODUCTION

Welcome to *Engineering for Cats*. To answer your first question: No, this is not a book about teaching cats how to be engineers. Although research has shown that cats actually may have enough brain capacity to perform rudimentary engineering methods,* they simply do not care enough to try (as with most activities). Instead, this book includes a series of projects designed to improve the quality of life for your cat or cats. Each project also provides some background engineering information on how it was designed or how it works.

---

* Not that cats are exceptionally intelligent. The bar to be an engineer is just quite low. We engineers use complicated-looking symbols to confuse our employers and convince them we're irreplaceable.

Some Practical Information on the Clientele

Cats are peculiar animals. People who love them can be just as confused by them as people who hate them. One of the most important things to understand about cats is that they come from a line of incredibly successful predators, and they are really just barely domesticated. The domestication process for most animals usually looks something like this:

Human recognizes utility of wild animal.

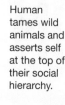

Human tames wild animals and asserts self at the top of their social hierarchy.

Human selectively breeds animal to amplify favorable traits.

Cats, on the other hand, have had a unique domestication process that went something like this:

Cat recognizes utility of living with human.

Human thinks this is adorable.

Human repeatedly attempts to find practical uses for cat with minimal to no success.

Cat is still adorable, so relationship continues.

It is often said that cats domesticated themselves because they were much more involved in the domestication process than any other animal. When human civilizations began to transition from nomadic groups to established cities, we inadvertently created a very favorable environment for many animals. Our trash (and the pests associated with it) provided cats with easier meals than they could get in the wild. The shelter from the environment probably seemed pretty attractive as well. Cats started hanging around more, and there was not a whole lot of motivation to drive them away. Major efforts to selectively breed them to make better pets were rare and ineffective.

In contrast, dogs were domesticated well before humans formed permanent settlements. The original utility of dogs was probably hunting or security, but they have been successfully bred to take on hundreds of specific jobs. Their domestication was about a productive and interactive partnership. But cats do not respond to training in the same way that dogs do. There is no social hierarchy for us to hijack (normally a key step of domestication), so training them doesn't fully work. The domestication of cats was really more about mutual tolerance.

Because humans had much less direct influence through the generations of cats,

our feline friends have retained more of their wild instincts than their cute little faces might indicate. Some of these instincts are what make them great companions. They are self-cleaning. They use a litter box. And they occasionally help get rid of pests such as rats or your lazy—ahem, allergic—ex-boyfriend. These traits, along with their entertainment value, are why cats are currently the most popular house pets in America, with an estimated count closing in on 100 million.

They have also retained some other instincts, which are often the source of household issues. House cats are still very territorial—in order for a cat to feel comfortable, she needs to be the master of her environment. This means being familiar with the layout of the space, not feeling threatened by other animals, and not feeling vulnerable while out in the open. Many of the projects in this book are aimed at making your cat feel like she has more control of her environment. These projects can be especially useful if you have multiple cats. Many people feel that their cat will be happier with another cat companion. Some shelters even push this idea to get you to adopt multiple cats. While there are plenty of benefits to having multiple cats, having to share their realm can

make them a little more defensive. This can lead to minor issues such as being less social, or more major issues involving inappropriate use of claws and pee. Adjusting the environment so that your cats are more comfortable is one way to address these issues.

House cats have also retained some level of hunting instinct. The majority of the time this just results in pure entertainment. However, if the urge to hunt is repressed, you can end up with a bored and overly energetic cat who happens to have daggers for nails. Many of the projects in this book are aimed at making this energy less destructive. While cats in any household can benefit from some extra stimulation, these will be especially helpful for indoor cats with a relatively small territory, such as an apartment.

An additional concept that is useful for understanding cat behavior is called a socialization period—the time window during which a kitten should be handled by humans

in order to feel comfortable around them, usually between two and seven weeks old. Experiments have shown that kittens who are not handled until after this period will likely be anxious around humans for the rest of their lives. (This is in contrast to puppies, who have a longer familiarization window and can eventually learn to relax around people with proper training if that window is missed.) This does not mean that a reclusive cat will not make for a great companion—there are other genetic and environmental factors that contribute to the personality of a cat. But these cats won't get stressed out if you don't give them enough attention, and when they do decide to sit on your lap it will be that much more special. The reason I mention the socialization concept is to acknowledge that some cat traits just cannot be changed. The projects in this book can help resolve behavioral issues and make your cats happier, but they are unlikely to change their overall personality.

Cat behavior is a relatively new area of study, but it has been increasing rapidly over the last century. If you are interested in getting a more in-depth understanding of your feline housemate, I recommend checking out the latest literature. Learning more about how we are essentially sharing our homes with self-domesticated natural predators who would probably eat you if you were small enough just makes the experience more enjoyable.

ENGINEERING FUNDAMENTALS
(OR, WHY THESE PROJECTS WILL WORK)

While the construction of these projects is not terribly technical, the design and function of each one is based on fundamental engineering concepts. It is not necessary to understand these concepts in order to build each project. In fact, your cat will probably not even know the difference if you skip these sections altogether—the project should turn out the same either way. The reason they are included is because they are universal concepts that you might find useful. The physics that apply to a cat structure are the same physics that apply to any other structure from the International Space Station to the Great Pyramids of Giza (although the pyramids might qualify as a cat structure anyway). Am I implying that by reading this book and building these cat projects you will be qualified to work on space stations? Absolutely. In fact, I would go so far as to claim that this book will teach you more about building space stations than any other do-it-yourself cat project book on the market, and I recommend listing any of the projects you do complete on your résumé. Feel free to list me as a reference.

BEFORE YOU BUILD

The information in this section is meant to reduce headaches and frustration as much as possible when constructing any of the projects. If you are fully capable in a workshop, then this information will probably not be new to you. If any of the information is new to you and you choose to learn it here, you should actually feel smart, because most of the pompous handymen and -women who will skip this section had to learn all of these tricks the hard way.

ACCURATE MEASUREMENTS

Everyone has heard the building advice "measure twice, cut once," but surprisingly few people know how to implement this rule correctly. Measuring twice does not mean taking your measurement and then taking the same measurement again. If you want to be sure that you are making a cut in the right place, you need to take two different measurements and confirm that they agree with each other. For example, say you want to cut a 96" piece of wood in half. First you measure 48" from one end and mark it. Then measure 48" from the other end and mark it. Only if your marks match up with each other have you successfully "measured twice."

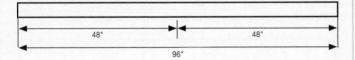

If you just repeat the same measurement from one end two times, then you will also repeat any mistakes you made and you haven't confirmed anything. I usually find a new way to mess up a measurement with every project I build, and if I did not follow this method I would be wasting a lot more material than I already do.

Cut to Fit

The projects in this book involve assembling multiple parts, a process that inevitably makes dimensions start to drift away from how they were drawn up on paper. Each time you add a piece, the errors can compound. This is not really a problem as long as you make adjustments along the way. If you are cutting a part to fit into an existing assembly, do not cut it based on the dimensions that are listed in the instructions—measure your existing assembly and cut it based on which dimensions are needed. The ideal dimensions are included as a reference. They can help you identify if something has gone horribly wrong.

If someone is looking over your shoulder and notices that you are re-measuring to fit in your next part, you can remind them that this is a result of manufacturing errors in the material itself. In wood, especially, there will be some warping and the dimensions will not be perfect, as trees stubbornly refuse to grow to perfect specifications. So the imperfections are a result of the quality of the materials you can get. They do not reflect on your skills as a craftsperson.

Mapping Cuts

Mapping out a large and complex cut can be much easier if you section off the profile and just copy it piece by piece. For example, one of the Advanced Cat Shelves (see page 12) has a profile approximately 2' × 4' in dimension that looks like this:

Freehanding that entire design onto a piece of plywood would be very challenging, and taking dozens of measurements to line it up would be tedious. To accurately transfer the profile, first draw a grid over it like this:

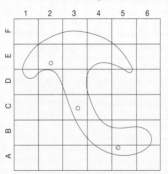

Draw a scaled-up grid on your plywood, then transfer the design square by square. Each section is simple enough to freehand, and you can reference the adjacent squares to see where the lines should cross into them. This is the easiest way to transfer a complex profile onto a piece of wood. We are assuming that you have a few basic tools on hand—such as pencils or pens, and a ruler or carpenter's square—to measure and mark cuts.

Scaling the Projects

When I adopted the two cats I have now, they were tiny kittens who could fit in the palm of my hand—maybe one pound each. I knew they would grow up eventually, but for some reason they just kept growing long after I expected them to stop. Now they are seventeen and twenty pounds, which my vet tells me is "impressive." These pudgy little tanks are the models for the projects in this book, and they are able to use all of them comfortably. This should mean that the projects will work for pretty much any size house cat. If you have a more average-size cat, you can scale some of the dimensions down a bit. Scaling them down would make them less expensive, easier to build, and more space efficient, so I recommend evaluating what the best dimensions are for you and your cat before diving in.

If your cat is much heavier than twenty pounds—first of all, congratulations. Your cat is impressively large and you may want to consider registering it with your local wildlife agency in the event that it ever escapes and wreaks havoc on your town. I would also urge caution if you intend to scale any of these projects up, especially with the load-bearing projects such as the shelves or wheel. The forces within the structure will not scale linearly, so you may have to evaluate the connections to make sure they do not break under the weight of your jungle cat.

HOW TO READ THE DRAWINGS

1. Approximate Cost: This is the cost assuming you have no starting materials at all. If you or your neighbor has a reasonably well-stocked garage, many of the materials needed should be on hand already and the total cost will be reduced. An exact cost is difficult to report because it can vary greatly depending on where you live and the decade in which you are currently living. If you notice a major difference in price at your local store, simply take this book to the register and explain that the purchase is for the benefit of a cat. If they do not discount your items, consider releasing hundreds of mice in the store late at night. To give some indication of what you'll spend, each project is labeled as follows:

Low: Less than or equal to the cost of a large bag of regular cat food (or a medium-size bag of the fancy stuff).

Medium: More than a large bag of cat food, less than the cost of a regular checkup at the vet.

High: More expensive than a regular checkup. The most expensive project in this book, the Cat Wheel, is about two to three times the cost of a checkup (but is still less than half the cost of wheels you can buy ready-made).

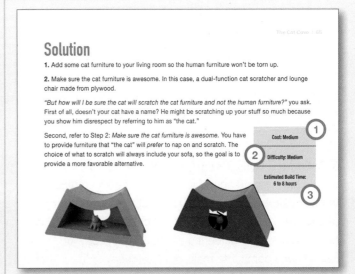

2. Difficulty: This indicates what experience level may be needed for each project. The levels of difficulty are:

Low: You'll still have to be willing to take some measurements and make some cuts, but you should be able to work your way through these projects. Even if you're the kind of person who routinely swears at inanimate objects for not cooperating.

Medium: I doubt anybody diligent enough to read this preface should be intimidated by a medium rating, so just go for it. I believe in you. Those people who skipped this section will probably make a royal mess of things, but you're better than they are.

High: Even the most difficult projects in this book are intended to be buildable by beginners, but if you are not experienced at making precise measurements and angled cuts you might want to buy some extra materials up front, just in case you need to take a few mulligans.

3. Estimated Build Time: This is my best guess. You should have a pretty good idea of what you're getting into, but it should be assumed that this estimate has a margin of error of +/- 100 hours, which is actually a pretty precise timescale for an engineering estimate. At my day job, I won't even commit to making a photocopy in fewer than three days.

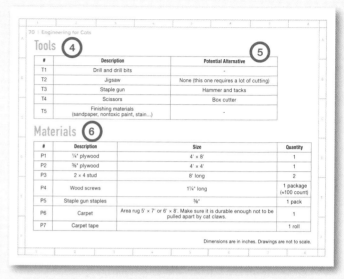

70 | Engineering for Cats

Tools ④ ⑤

#	Description	Potential Alternative
T1	Drill and drill bits	-
T2	Jigsaw	None (this one requires a lot of cutting)
T3	Staple gun	Hammer and tacks
T4	Scissors	Box cutter
T5	Finishing materials (sandpaper, nontoxic paint, stain...)	-

Materials ⑥

#	Description	Size	Quantity
P1	¼" plywood	4' × 8'	1
P2	⅜" plywood	4' × 4'	1
P3	2 × 4 stud	8' long	2
P4	Wood screws	1¼" long	1 package (≈100 count)
P5	Staple gun staples	⅜"	1 pack
P6	Carpet	Area rug 5' × 7' or 6' × 8'. Make sure it is durable enough not to be pulled apart by cat claws.	
P7	Carpet tape		1 roll

Dimensions are in inches. Drawings are not to scale.

4. Tool List: All of the projects are meant to require common tools that every garage should have, but not everyone has the same idea about what a garage should have in it. For example, some people will store a car in their garage despite the fact that it is basically useless for woodworking and tends to just get in the way. It is possible that these lunatics will also want to build elegant projects for their cats, so I have included a list of needed tools to avoid excluding them.

5. Potential Alternative: The first tool listed is usually the preferred or the simplest option, but there are always many different tools that can produce the desired result. Such alternatives are listed here, but you should always read through the instructions in full and plan out the tools you need based on what you have available and what you are most comfortable using. I have tried to list the simplest versions of the tools that will work, so if you have a more sophisticated workshop you can probably identify ways to make these projects much more easily.

6. Materials List: Unless otherwise indicated, all of these materials should be available at any hardware store.

7. Instruction Pages: The subsequent pages have a step-by-step guide for constructing the project. Make sure to read them all the way through before diving in.

8. Part Multiplier: Some parts need to have multiples made, as indicated by this type of note.

9. Tools/Materials Box: These boxes reference which tool or raw material will be required to complete the steps on each page. They do not include a few basics, such as pencils and a carpenter's square or ruler, that you should already have on hand to measure and mark cuts.

WORKING WITH WOOD

The most utilized building material in this book is wood. It's easy to work with, inexpensive, and a good way to assert dominance over trees. There are many qualities of wood from which to choose, each having its own benefits and drawbacks for different projects. The lower qualities will have more defects such as knots and splits, less precise dimensions, and the surface will not be as finished as the pricier options.

These factors will not be issues if you choose the pieces that you buy carefully and properly finish the wood yourself.

Before you buy a piece of wood, make sure to check that it is straight, free from warping, and does not have any major splits or knots. For many large construction projects

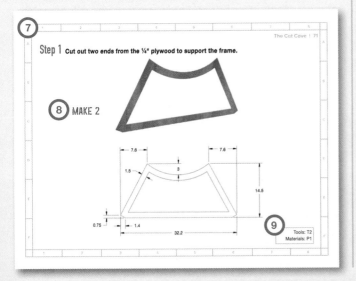

The Cat Cave | 71

Step 1 Cut out two ends from the ¼" plywood to support the frame.

MAKE 2

7.6   7.6
1.5   3
14.5
0.75   1.4
32.2

Tools: T2
Materials: P1

these defects would be inconsequential, but the projects in this book will come out much better if you choose carefully. If you spend a little extra on higher quality timber, you can save some time, but keep in mind that this will also rob you of the enjoyable time you get to spend finishing the wood (for those who take pleasure in the process).

Finishing

The more time you put into finishing wood projects through sanding, sealing, and painting, the better the final product will look. At minimum you should always give every wood project a few minutes of sanding. Breaking any sharp edges and removing splinters will definitely be required. Sandpaper is a required tool for every woodworking project.

Pre-drilling

When a screw is added to a piece of wood, the surrounding wood has to expand just a little bit to make room for the material of the screw. If the screw is too close to an edge and the wood is too brittle, adding a screw can crack the wood. To avoid this damage, pre-drill a small hole to remove some material in advance so that the screw doesn't have to push as much out of the way. In the instructions for each project I try to indicate wherever pre-drilling might be necessary, but whether or not this is necessary depends on the skill of the operator and the quality of the wood. So

if you do crack a part while adding a screw, just blame the wood. This excuse may be difficult to use depending on where you're from. I don't know where you might be getting your lumber, but here in America our high-quality timber comes from a place where the majestic trees have grown strong for generations and won't easily give way to the torment of a wood screw: Canada.

Many skills required in these projects are not completely spelled out. Skills such as evenly applying stain, drilling a straight hole, or making an accurate cut are critical to the success of these projects, yet I have ruthlessly left you high and dry. Aside from my irrational fear that you will complete better versions of these projects than I have, some skills have been omitted because they depend on what materials and tools you choose to use. Fortunately, the instructions for these materials (such as stain) or tools (such as a jigsaw) are usually found on the packaging. The people who wrote those instructions know the material much better than you or I, so follow them carefully.

Don't worry if you mess up—it is unreasonable to assume things will go perfectly and no amount of text could have made it so. The one exception to learning through experience is in safety gear. Learning how to prevent personal injuries by injuring yourself is not ideal. Never rush, and never use tools improperly.

SIMPLE CAT SHELF

Problem

Cats will employ destructive behavior to gain a height advantage.

Elevating Good Behaviors

Most cats see themselves as, at the very least, equal to their humans—a fact that most cat owners can attest to. Situations that may make cats feel inferior can result in their overcompensating with negative behaviors. For example, because cats are typically only about one foot tall and even fledgling humans are at least two, a cat may feel a bit vertically challenged. When humans surrender their height advantage by lying down, many cats are ready to take advantage of the situation (SEE FIGURE 1). Other cats may develop impressive tripping strategies (SEE FIGURE 2) when they see you carrying anything bulky.

Figure 1.
Cats are well known for being opportunistic hunters.

Figure 2.
When cats recognize your vulnerability—limited field of vision and raised center of gravity—they may take advantage of the situation.

Solution

A simple yet reliable shelf that your cat can use to gain a higher vantage point, or as a step to an existing perch. This can help boost a cat's confidence, and make him feel more comfortable out in the open.

To soothe the ego of your cat and reduce her adorable yet destructive behavior, provide access to elevated perches throughout your house. A series of well-placed shelves will make your cats feel more comfortable in their domain (and may draw out those who are timid). The shelf design provided in this section can be used as an effective perch, or it can be used as a stepping-stone to hard-to-access spots. If you have more of a blank canvas (aka a big empty wall) and are looking for more of an aesthetically pleasing series of shelves, see the advanced shelf project in the next chapter.

Cost:
Low

Difficulty:
Low

Estimated Build Time:
1 to 2 hours

SECURING THE SHELF TO A WALL

In order to mount this shelf you will most likely have to work through drywall. Drywall is mostly made of compressed plaster, and from an engineering perspective it is considered nonstructural. It's a good material for sealing out the rest of the world and creating an isolated life with your cats, but it is not quite as good as a load-bearing surface.

Drywall is the most popular option for interior walls, but if your walls are made of another material, you might need specialty fasteners and drill bits. A quick internet search should reveal effective ways to mount this shelf to just about anything.

Finding a Stud Behind the Drywall

The wood studs behind the drywall are what actually provide the structural support of your house or apartment, and those studs are what you want to be drilling into. The challenge, though, is in locating them. They are typically spaced about 18" apart and should be 1½" wide, but there is really no reason why they would be spaced precisely, so locating them by measurement is not usually a viable option.

The most reliable way to locate a stud is with a stud detector (also called a stud finder). They are simple to use—simply move the stud detector over the wall slowly and wait for the device to blink or beep to tell you that it's over a stud—and can range from $10 to $100, but I do not recommend spending more than $25.

Quarter Tap Test

If you don't want to spend money on a stud detector, there is a cheaper option that costs only 25 cents. It is called the quarter tap test, and it is often used to identify damage on aerospace structures made of composite materials such as carbon fiber. The method works like this:

Repeatedly tap a quarter on the wall around the location where you are trying to identify a stud. Listen carefully to the pitch of the sound the tapping makes. When you tap on a location over a stud, the pitch will get higher. When you tap on a location that is not over a stud, the pitch will be lower.

The reason the pitch gets higher when tapping over

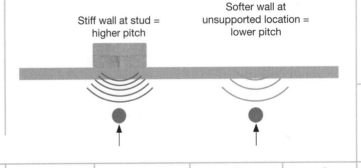

Stiff wall at stud = higher pitch

Softer wall at unsupported location = lower pitch

a stud is because the wall is stiffer. This method isn't foolproof, and it can require a keen ear and a little bit of practice, but if you are too cheap or lazy to go buy a stud detector, it can be the next best alternative.

You will know if you have successfully screwed into a stud if the screw pulls the shelf in tight against the wall once it is all the way in. If you miss, the screw may just continue to spin around after it is bottomed out or it just won't pull the shelf in tight.

Repairing Drywall

Previous laziness and frustration has led me to employ the "guess and check" strategy of locating a stud, which does eventually work but will leave your wall full of holes. Fortunately, these are very easy to repair by smearing some drywall spackle into the holes. Drywall spackle can usually be found in the paint section of any hardware store, and a bucket capable of erasing hundreds of holes is only about $5. If you want to make it look nice, you can sand down the dried spackle with fine grit (220+) sandpaper and then paint over it.

A common dorm room solution to repairing holes is to use toothpaste. Do not do this. Toothpaste is way more expensive than spackle and it doesn't stick to paint. So if the wall is ever painted again, all of your lovely mistakes will become embarrassingly obvious.

ENGINEERING 101
Dynamic Cat Loading

Why go to all this trouble to make a sturdy shelf? You might be able to find a wall-mounted bookshelf that says it is capable of holding 20 to 30 pounds, so if your cats weigh less than that, those shelves should work . . . right?

The answer is unfortunately no. Although your twenty-pound cat may equal the weight of twenty pounds of books, there is a small complication in that cats are *slightly* more mobile than books. Books on a shelf would be referred to as a *static load*, meaning it is applied slowly and doesn't move. A cat jumping on a shelf, however, is a *dynamic load*. When a cat is applied dynamically, the force can be amplified many times over.

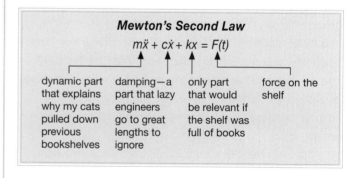

Mewton's Second Law

$$m\ddot{x} + c\dot{x} + kx = F(t)$$

| dynamic part that explains why my cats pulled down previous bookshelves | damping—a part that lazy engineers go to great lengths to ignore | only part that would be relevant if the shelf was full of books | force on the shelf |

This can be shown analytically using Newton's second law, one on which my engineering career is almost exclusively based. Here is a general explanation: If you placed a bowling ball on your foot, it shouldn't hurt that badly—it might just feel a little awkward. But if you dropped a bowling ball on your foot, even from just a foot or two, that could hurt. That's Isaac Newton's fault.

Static Load

Dynamic Load

Tools

#	Description	Potential Alternative
T1	Circular saw	Jigsaw, Handsaw
T2	Drill and drill bits	-
T3	Wood glue	-
T4	Stud finder	Quarter
T5	Sandpaper (The grit needed will depend on the quality of the wood. For best results, do a first pass with 80 and a final pass with 120 or higher.) and paint or stain, if desired.	-

Materials

#	Description	Size	Quantity
P1	½" plywood	2' × 2'	1
P2	2 × 2 wood stud	8' because that will probably be the only option; you will only need 16" of it, though	1
P3	Short wood screws	¾"–1¼" long	6
P4	Long wood screws	2"–3" long	4

Dimensions are in inches.
Drawings are not to scale.

Step 1 Use a pencil and a ruler or carpenter's square to mark the cuts shown below on the plywood.

The shelf should be wide enough to reach two wall studs. If it's not, adjust accordingly. Cut out the top, back, and two support pieces. Clean up the edges with sandpaper.

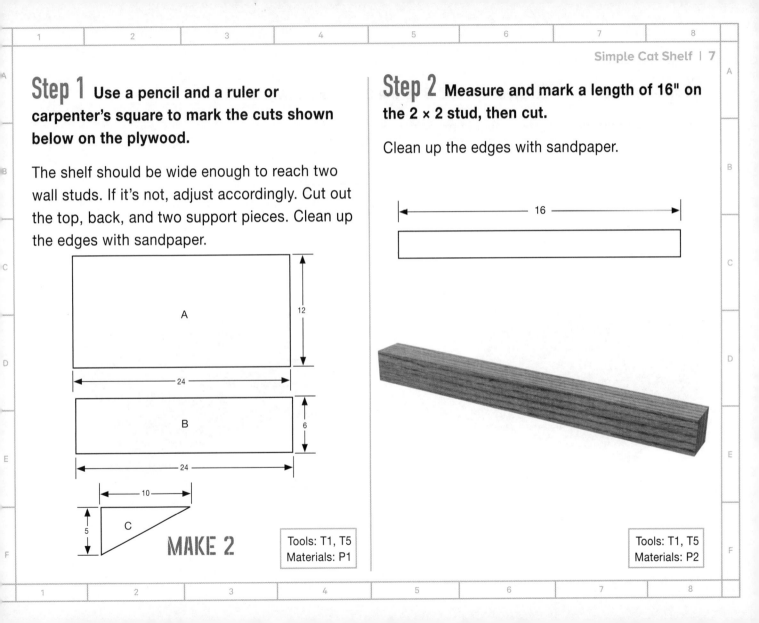

A

12

24

B

6

24

10

5

C

MAKE 2

Tools: T1, T5
Materials: P1

Step 2 Measure and mark a length of 16" on the 2 × 2 stud, then cut.

Clean up the edges with sandpaper.

16

Tools: T1, T5
Materials: P2

Step 3 Attach plywood pieces A and B to the 2 × 2 as shown.

Space the short wood screws evenly on each side.

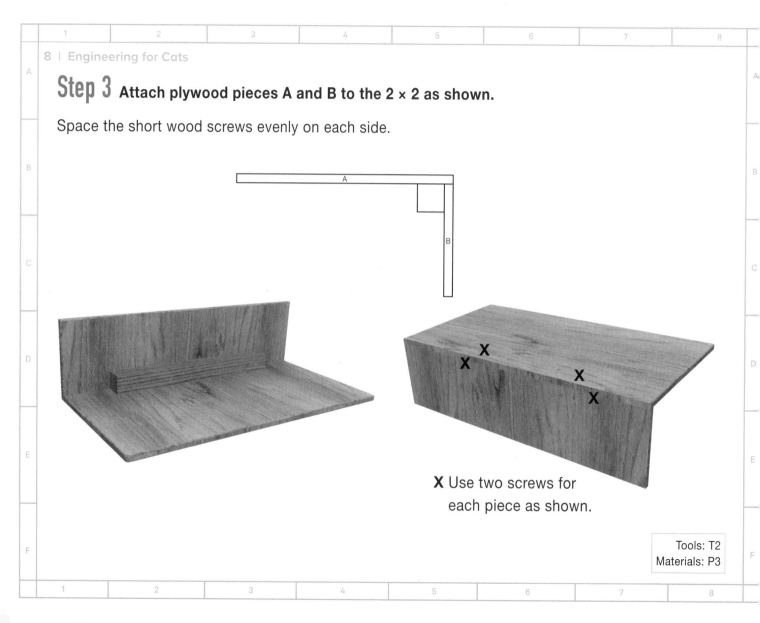

X Use two screws for
each piece as shown.

Tools: T2
Materials: P3

Step 4 Attach the two triangular pieces (C) as shown.

Use wood glue along the edges where they interface with pieces A and B. Add a small wood screw through each side, into the ends of the 2 × 2.

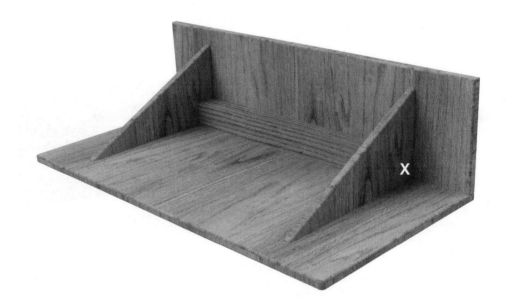

Tools: T2, T3
Materials: P2, P3

Step 5 Finish the shelf.

Sand as needed. For a natural wood look, you can stain the shelf and then seal it with polyurethane instead of painting it. Adequate nontoxic paint, stain, and polyurethane (a textured polyurethane might be helpful if your cat needs a little assistance with grip for landings and takeoffs) will be available in a standard hardware store. Follow the application instructions on each container.

Tools: T5

Step 6 Attach the shelf to the wall.

Line it up so that two long wood screws can anchor into each stud as shown. If attaching to drywall, follow the guide on p. 3.

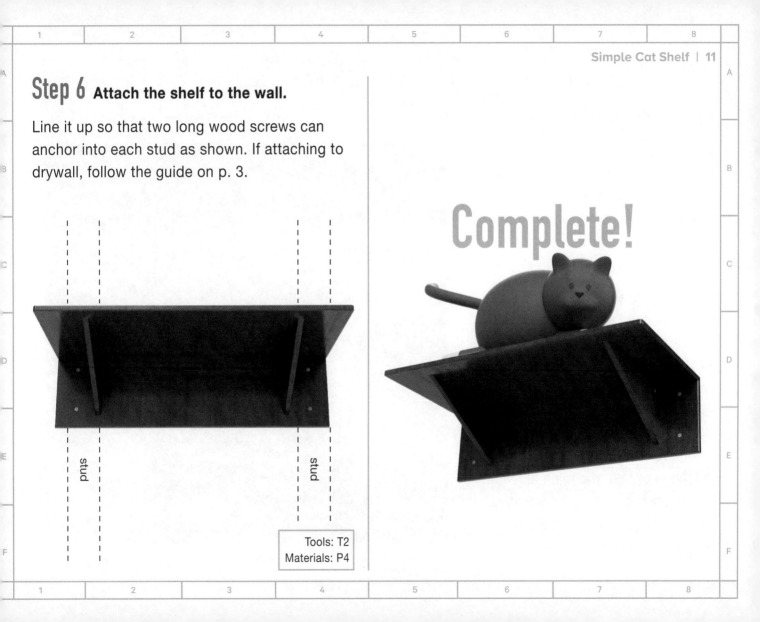

stud

stud

Tools: T2
Materials: P4

Complete!

ADVANCED CAT SHELF

Problem

Cats will (continue to) employ destructive behavior and will require yet more fancy shelving to gain a height advantage.

Elevated Expectations

The benefits of giving your cat the opportunity to live the high life are introduced in the previous project. This extended variation will present some shelving options that are a little more aesthetically pleasing and more comfortable for your cat to lounge on, but the main function is the same: providing an elevated perch to help your cat deal with its Napoleon complex.

Giving an opportunistic hunter a prime opportunity to hunt *you* should be done with caution.

Solution

Designed to be more attractive and more comfortable for cats than the standard shelf, a curved shelf will give cats a better vantage point.

To maximize comfort, these shelves (and a bonus—a ramp for the less agile) include a carpeted surface and a contoured profile. The carpet allows the shelf to be used as a scratching post, which can help your cat establish ownership of the perch. The curvature will make the shelf more ergonomic for lounging, which is fairly important as well. A cat lazing on a high shelf will look like he's just enjoying the view, but a cat perched alertly on a high shelf can look more like a scheming predator trying to size you up. The latter case more accurately represents what is going on in your cat's head, but reminders of this seem unnecessary.

Cost: Medium
Difficulty: Medium
Estimated Build Time: 4 to 6 hours

ENGINEERING 101

Force Reactions

The job of any static structure is to transfer a force into the ground. A chair distributes your weight between its legs and then into the ground; a bicycle transfers your weight through the frame, to the spokes of the wheel, into the tire, and then into the ground—a lot of different structures are desperately trying to pawn off your weight into the ground. If you're anything like my cats, then you have some excess weight to be distributed, meaning any supporting structure will have to be strategic about this transfer.

In addition to adding to the comfort and style of lounging, the curved shelf is a structural feature specifically designed to withstand heavy (but mobile) house lions.

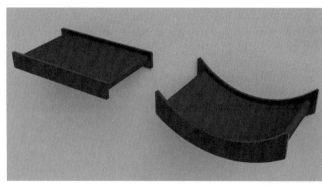

First, consider the flat shelf (below). When the force from the cat is applied to the center of the shelf (1) it is transferred into the wall. The top part of the shelf-to-wall interface will be pulled apart (2) while the bottom part will be squished together (3). The most likely way that this interface would fail would be for the fasteners on the top (2) to pull out of the wall.

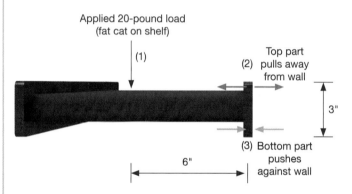

Applied 20-pound load
(fat cat on shelf)

(1)

Top part
(2) pulls away
from wall

(3) Bottom part
pushes
against wall

3"

6"

We can make a simplified calculation for how much force will be applied to this fastener (aka I can do it and you can read the answer). If the distance from the wall to where the twenty-pound cat load is being applied is 6", and the distance from the top fasteners to the bottom of the shelf is 3", then the force pulling the top fasteners away from the wall will be forty pounds. (For more information on this calculation, see page 15.)

Now let's take a look at the curved shelf (below):

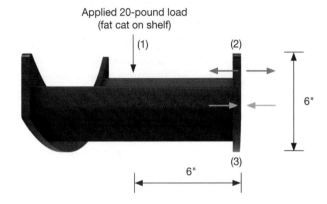

Applied 20-pound load
(fat cat on shelf)

(1)

(2)

(3)

6"

6"

The twenty-pound cat load is still being applied 6" away from the wall, but the top fasteners are now 6" away from the bottom of the shelf. In this shelf, the force pulling the top fasteners away from the wall will be only twenty pounds. By doubling the distance between the upper and lower fasteners, the load on the critical fasteners was cut in half (meaning your cat can get twice as fat!).

The minimized support structure of this "floating" shelf design leaves little room to add many fasteners, so we start off by using a few big ones called lag bolts. Lag bolts are just large sturdy screws, usually with a hex-style head so they can be torqued down firmly. They are used to create a solid attachment between the end face of the shelf and the support beams inside the shelf. There is then plenty of space to attach the backing plate into this end piece with normal wood screws. The backing plate will cover enough area that it can be attached through two different studs in the wall. This attachment is intended to provide enough strength to avoid failing, as well as enough stiffness to remain immobile when a cat jumps onto it.

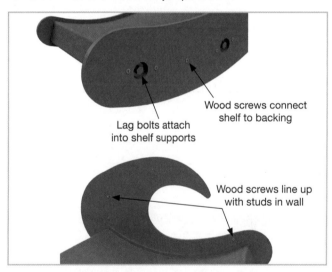

Wood screws connect shelf to backing

Lag bolts attach into shelf supports

Wood screws line up with studs in wall

Simplified Force Calculation

For those who are still reading, a little more detail on the fastener forces presented above may be of interest. First, let's come clean on some lies that were told. The calculation is based on a fairly simplified model of the shelves. In reality, the entire end of the shelf is in contact with the wall, so there is a distributed pressure over the bottom part that is pushing against the wall. For the calculation, this was simplified to assume that all of the load is being transferred at the fasteners. A simple diagram of this idealized shelf* is shown below:

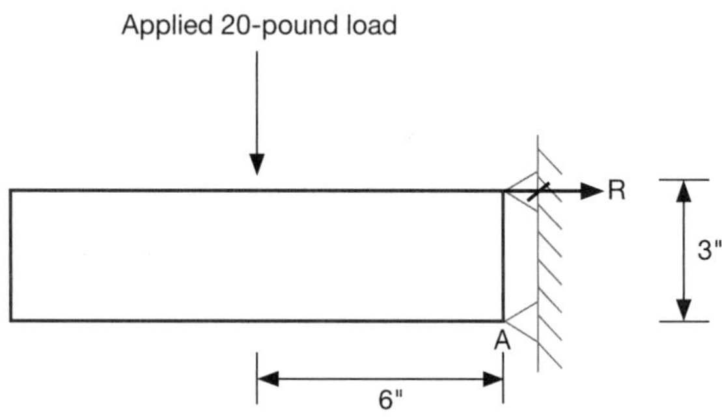

Applied 20-pound load

R

3"

A

6"

From this idealized model, calculating the force pulling out the top fasteners is actually pretty simple. Because the shelf is not supposed to move, we assume all of the forces acting on it are in equilibrium, meaning they will sum up to zero. In this case, we can sum the "moment" forces about

---

* This is what the shelf would look like if all the forces were behaving exactly as they have been represented in the calculations.

point A as shown in the diagram. A *moment* is a rotational force defined as the force multiplied by the distance away from the pivot point where it is being measured (referred to as the "moment arm"). More details about moments are included in the Drawbridge Cat Door (see page 129). The only two forces that are trying to impart a rotation about point A are the weight of the cat, which is applied 6" away, and the reaction force in the top fasteners, which is applied 3" away ("R" as shown at left). They act in opposite directions about point A, so we can consider one a positive moment and one a negative. Combine these moments and set the result equal to zero:

$$20(6) - R(3) = 0$$

Use algebra to solve for the reaction force:

$$R = 20(6)/3$$

. . . and we can see that R = 40. If the fasteners are set farther apart, the reaction force will go down, and this is one advantage to having a curved shelf. The actual force is slightly different due to the simplified boundary conditions (the interface at the wall) we assumed, but the point—to show how spreading out the fasteners requires them to handle less force—is still true.

Advanced Shelf

Tools

#	Description	Potential Alternative
T1	Drill and drill bits (up to 1" bit)	-
T2	Jigsaw	-
T3	Scissors	Box cutter
T4	Staple gun and staples	Hammer and tacks
T5	Finishing materials (sandpaper, and paint or stain, if desired)	-
T6	Wrench	-
T7	Stud finder	Quarter

Materials

#	Description	Size	Quantity
P1	¼" plywood	4' × 8'	1
P2	½" plywood	2' × 2' (see Step 6)	1
P3	2 × 4 beam	8' long	1
P4	Short wood screws	¾" to 1¼" long	1 package (≈100 count)
P5	Long wood screws	2"–3" long	6
P6	Carpet	A cheap area rug approximately 4' × 6' will probably be the most cost-effective. This will leave you with some extra carpet for additional shelves.	1
P7	Carpet tape	2" wide	1 roll
P8	Lag bolts	½" to ⅝", ≈3" long	3
P9	Flat washers	Large enough for lag bolts	3

Dimensions are in inches.
Drawings are not to scale.

Step 1 Cut two end pieces for the frame out of ¼" plywood.

Dimensions are approximate and can be tailored based on the size of your cat.

MAKE 2

R18

1.5

21.7

Tools: T2
Materials: P1

Step 2 Cut out the support beams.

Measure, mark, and cut three pieces, each 12" long, from the 2 × 4, as shown.

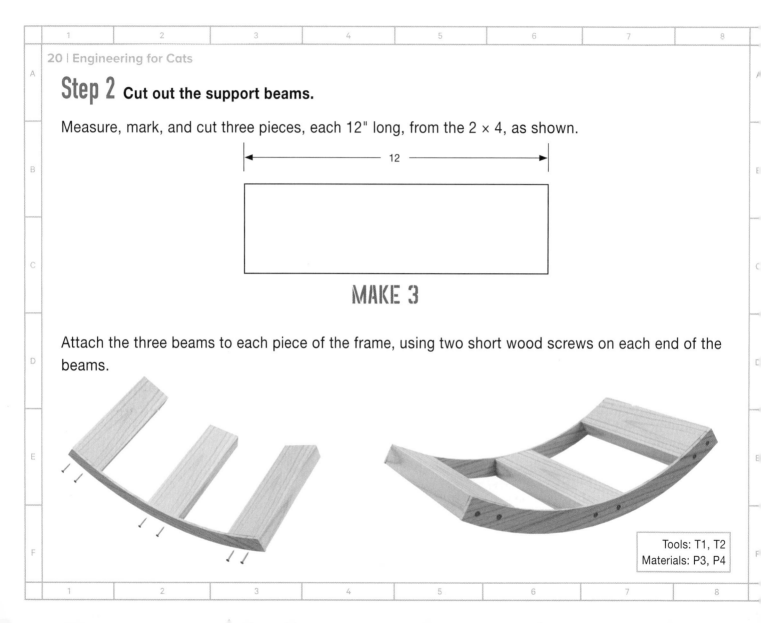

|← ———— 12 ————— →|

MAKE 3

Attach the three beams to each piece of the frame, using two short wood screws on each end of the beams.

Tools: T1, T2
Materials: P3, P4

Step 3 Cut the top and bottom panels from the ¼" plywood to the dimensions shown.

Attach the panels to each beam with three short wood screws (nine screws total per panel).

For more information on getting the wood to conform to a curve, see pages 67–68.

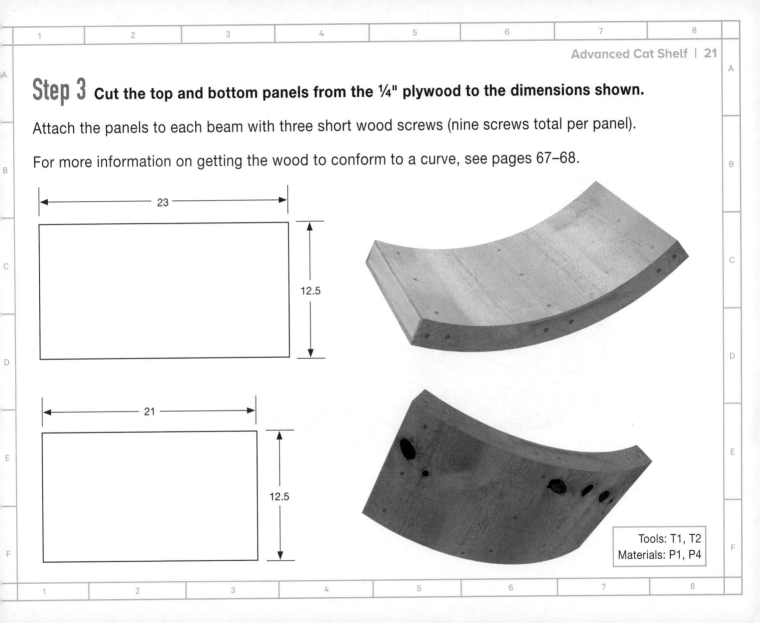

23

12.5

21

12.5

Tools: T1, T2
Materials: P1, P4

Step 4 **Wrap the constructed arc in carpet.**

Use plenty of carpet tape to secure the carpet to the top and bottom faces of the shelf. Staples or tacks should be needed only on the sides.

You will have to cut the carpet strategically to fold it over the edges of the arc. See the Cat Cave project on page 64 for more detailed carpet-wrapping advice.

Tools: T3, T4
Materials: P6, P7

Step 5 Cut out, paint, and attach the end pieces.

Cut two end pieces from ¼" plywood, approximately ½" larger than the arc on all sides, with a rounded end.

Sand and paint before attaching to each end with wood screws as shown.

Attach the front end piece with short screws to the support beams as shown; attach the back end piece with just a few screws to hold in place until the lag bolts are added (see Step 7).

MAKE 2

0.5

Attach the end pieces to each support beam with wood screws.

Tools: T1, T2, T5
Materials: P1, P4

Step 6 Cut out the backing piece from ½" plywood.

The shape of this piece can be whatever design you choose, as long as it has enough area to get several fasteners into two different studs on the wall. Maintain at least a 1" gap around the arc end piece.

Add three holes centered at the support beams in the backing plate as shown. These will allow access for lag bolts to create a secure connection to the arc (see Step 7).

Sand and paint before the next step.

Locate the holes centered over the support beams in the arc. The holes have to be just large enough to make room for the flat washers (P9).

Tools: T1, T2, T5
Materials: P2

Sample inspiration for other backing pieces:

Giant Cat Face

Sailboat

Classic Wood Finish

Island

Mt. Olympus

Step 7 Use wood screws to attach the backing piece to the arc and add lag bolts.

Add screws on either side of each hole for the lag bolts, and also in spaces between them as shown.

Pre-drill holes into the arc's support beams and install lag bolts with flat washers.

Tools: T1, T6
Materials: P4, P8, P9

Step 8 Screw the backing piece into the wall.

For the shelf to be secure, it must be installed into two solid wall studs. Use longer screws (2" or 3" long) and make sure they are secured in the studs. If they continue spinning after they pull all the way in, they are not properly seated.

See the Simple Shelf instructions (page 1) for further advice on mounting shelves.

stud

stud

Tools: T1, T7
Materials: P5

Complete!

Advanced Shelf Ramp

This shelf is designed to be a complementary piece to the curved shelf. It can be used as intended, or oriented horizontally to create a longer single shelf.

Cost: High

Difficulty: Medium

Estimated Build Time:
5 to7 hours

Tools

#	Description	Potential Alternative
T1	Drill and drill bits (up to 1" bit)	-
T2	Jigsaw	-
T3	Scissors	Box cutter
T4	Staple gun and staples	Hammer and tacks
T5	Finishing materials (sandpaper, and paint or stain, if desired)	-
T6	Wrench	-
T7	Stud finder	Quarter

Materials

#	Description	Size	Quantity
P1	¼" plywood	4' × 8'	1
P2	½" plywood	4' × 8' (see Step 6)	1
P3	2 × 4 beam	8' long	1
P4	Short wood screws	¾" to 1¼" long	1 package (≈100 count)
P5	Long wood screws	2"–3" long	6
P6	Carpet	A cheap area rug approximately 4' × 6' will probably be the most cost-effective. This will leave you with some extra carpet for additional shelves.	1
P7	Carpet tape	2" wide	1 roll
P8	Lag bolts	½" to ⅝", ≈3" long	3
P9	Flat washers	Large enough for lag bolts	3

Dimensions are in inches.
Drawings are not to scale.

Step 1 Measure, mark, and cut two end pieces for the frame as shown, out of ¼" plywood.

Dimensions are approximate; they can be tailored to the space available on the wall.

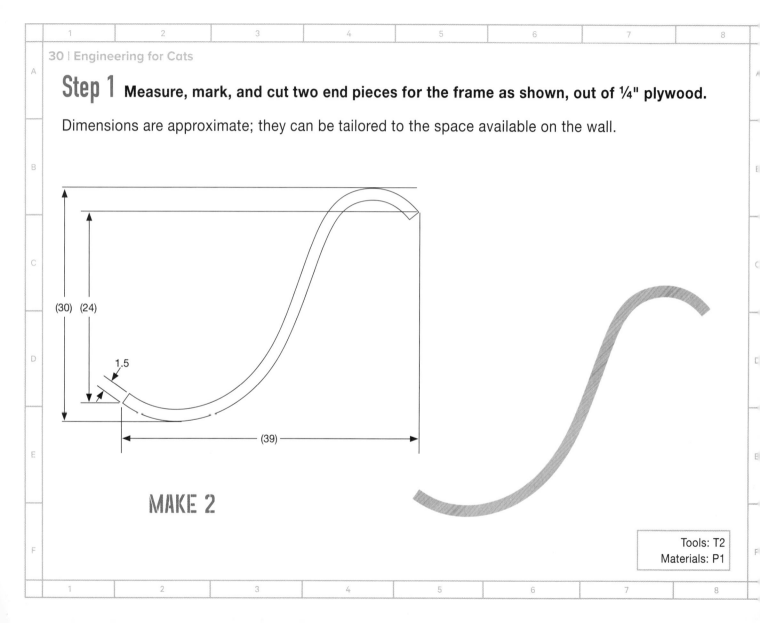

(30) (24)

1.5

(39)

MAKE 2

Tools: T2
Materials: P1

Step 2 Attach support beams.

Measure, mark, and cut seven pieces each from the 2 × 4, 9" long, as shown. Attach the pieces to the frame with two short wood screws on each end.

MAKE 7

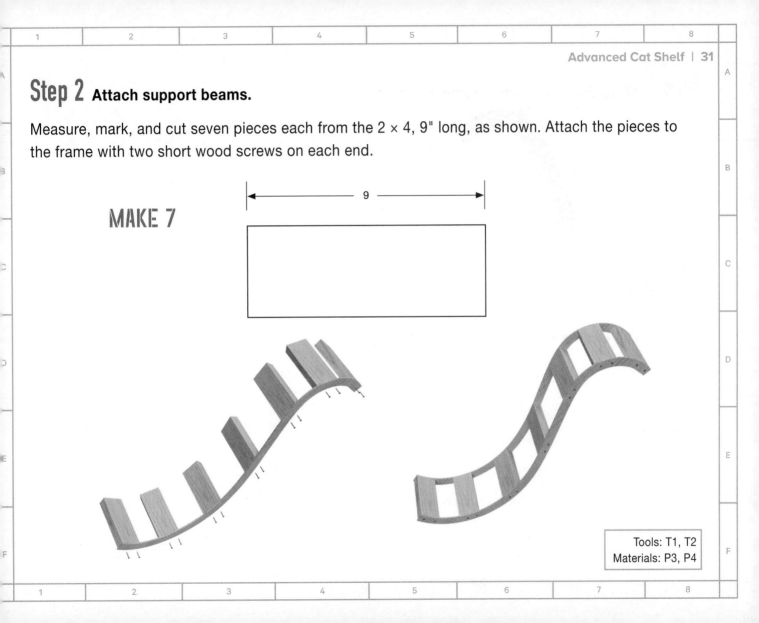

Tools: T1, T2
Materials: P3, P4

Step 3 **Wrap the constructed frame in carpet.**

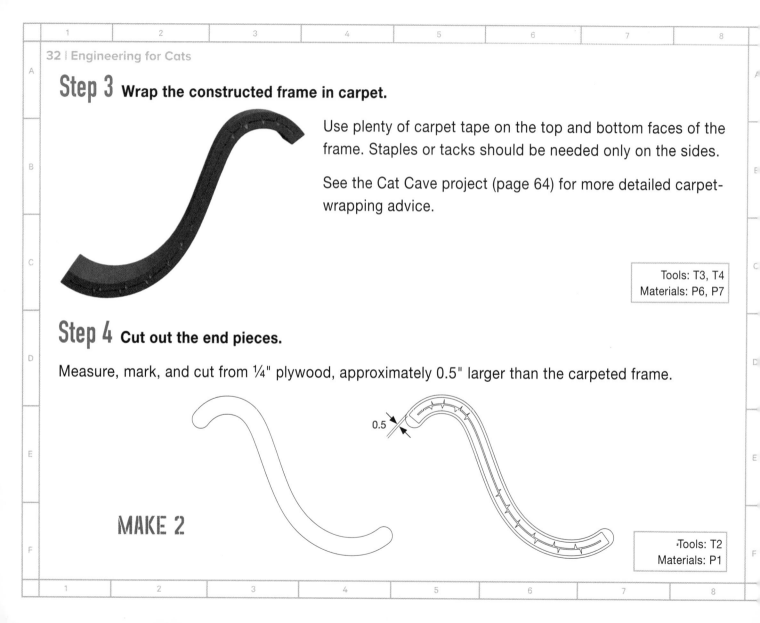

Use plenty of carpet tape on the top and bottom faces of the frame. Staples or tacks should be needed only on the sides.

See the Cat Cave project (page 64) for more detailed carpet-wrapping advice.

Tools: T3, T4
Materials: P6, P7

Step 4 **Cut out the end pieces.**

Measure, mark, and cut from ¼" plywood, approximately 0.5" larger than the carpeted frame.

0.5

MAKE 2

Tools: T2
Materials: P1

Step 5 Paint and attach the end pieces.

Attach the front end piece with short wood screws into the support beams as shown. Attach the back end piece with just a few short wood screws to hold in place until Step 7, when you add the lag bolts.

Tools: T1, T5
Materials: P4

Step 6 Cut the backing piece out of the ½" plywood.

The shape can be whatever design you choose, as long as it has enough area to get several fasteners into two different studs on the wall (see page xv for instructions on how to make a grid to transfer your design onto the plywood).

Add three holes centered in the backing plate as shown. The holes should be centered on the 2 x 4 support beams from Step 2, and large enough to make room for flat washers. They will allow access for the lag bolts to create a secure connection to the arc (see Step 7).

Paint before the next step.

0.5

Maintain at least a ½" gap around the arc end piece.

Orient the holes so that they are centered over the support beams in the arc.

Tools: T1, T5
Materials: P2

Step 7 Use short wood screws to attach the backing piece to the arc; add lag bolts to secure the connection to the frame.

Add screws on either side of each lag bolt and in spaces between them as shown.

Pre-drill holes in the arc and install lag bolts with flat washers.

Tools: T1, T6
Materials: P4, P8, P9

Step 8 Screw the backing piece into the wall.

For the ramp to be secure it must be installed into two solid studs. Use long wood screws and make sure they catch in the studs. If they continue spinning after they pull all the way in, the screws are not properly seated. Add one screw below the ramp surface and space out two or three above it. See the Simple Cat Shelf section (page 1) for further advice on mounting shelves.

stud

stud

Tools: T1, T7
Materials: P5

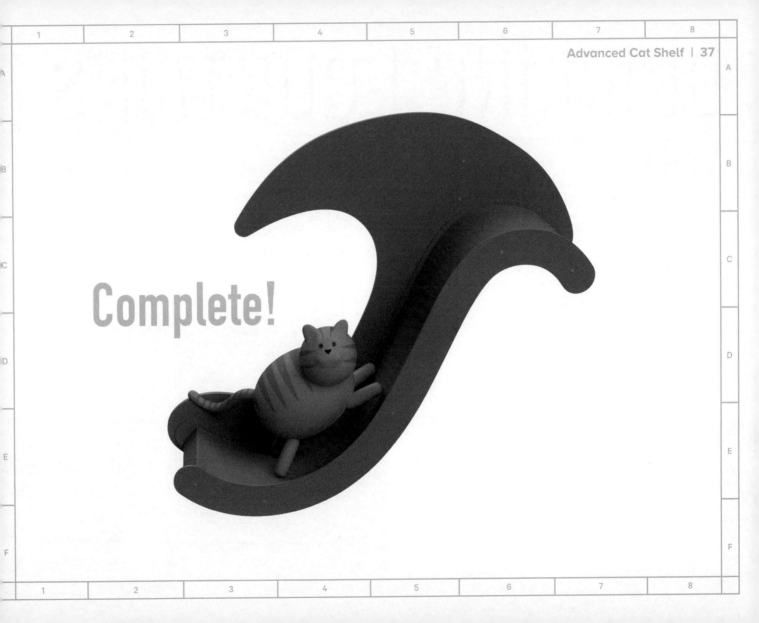

Complete!

DRINKING FOUNTAINS

Problem

Your cat may be dehydrated and cranky.

Hyper-Carnivore Hydration

House cats come from a line of "hyper-carnivores" that do not actually need too much water. They can get almost all the hydration they need from their food . . . provided that the food consists of freshly murdered woodland creatures. If your cat's diet consists of dry food, she'll need some supplemental water. If she is given dry food but doesn't ever seem to need water, then I've got some news for you about what your vicious little companion might be doing in her free time.

Solution

Provide moving water with a fountain.

Many owners notice that their cats will go crazy for water out of a running faucet. This instinct could be natural, because stagnant water in the wild is more likely to be contaminated than running water. Whatever the reason, giving them a recirculating water fountain to drink from can help them drink more without wasting. The first fountain (small, p. 43) is an inexpensive and easy-to-build option designed to be placed in a water dish. Cats can drink off the top, from the front, or just from the bowl. The second option (large, p. 52) provides more drinking locations, can include a water filter, and is dishwasher-safe.

It is important to keep in mind that any fountain used for drinking (even a store-bought one) will need to be cleaned occasionally. This fountain can be easily split into two pieces and run through dishwasher. I recommend making two and switching them out to keep them clean (the same reason why I own two forks). You should only have to switch them out about every week or so, but it will depend on how often your cat sticks its furry face in this thing.

Note: If your cat enjoys the thrill of potential electrocution via chewing cords, you might want to reconsider making this project.

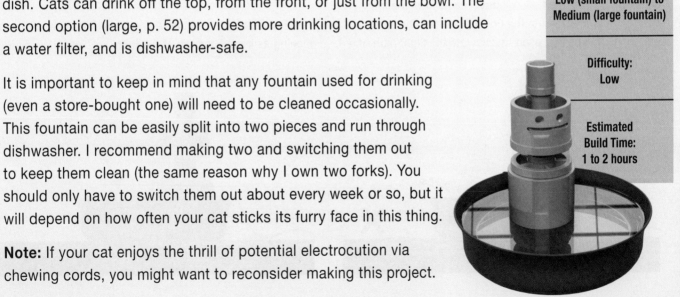

Cost:
Low (small fountain) to
Medium (large fountain)

Difficulty:
Low

Estimated
Build Time:
1 to 2 hours

ENGINEERING 101
Splash Mechanics

Projectile Motion

As cool as aerial water features can be, we want to minimize splashing as much as possible in this fountain. Any splashing will slowly but consistently moisten everything around it, to the displeasure of you, your cat, and potentially any downstairs neighbors.

Using the equations of projectile motion, we can calculate how far the flow will travel away from the fountain before it hits the water in the bowl. These are the same equations that can be used to calculate the flight path of a cannonball fired from a cannon, a cannonball dropped off a cliff, or a cannonball sitting on the ground not moving. (Basically, before Sir Isaac Newton developed these equations, there was a lot of mystery associated with cannonballs. Appreciate the times you live in.)

We need to know the horizontal velocity of the water as it exits the fountain. The pump I recommend outputs water at a rate of 40 gallons per hour, but the rate at which it comes out of the fountain depends on how much area the water has to squeeze out of. Specifically:

water exit velocity = (pump flow rate) / (exit area)

The smaller the exit area, the faster the water will be moving as it exits the top of the fountain. This is just like

SPLASH ✗

✓

putting your thumb partially over the end of a hose to spray the water farther. You can design the top of this pump in many different ways (see Step 5), but to keep the water from splashing out you need the total exit area to be large enough to slow things down.

We want the water to travel no more than ¼" away from the fountain, which will require a total exit area of at least 1.4 in². The pump should have a control switch to limit the flow rate. If you set the pump at half capacity to allow room for adjustment, it should pump at a rate of 20 gallons per hour. At this rate you would need the exit area to be 0.7 in². As a reference, if you plan on drilling holes to create the exit area, you would need to drill about 2 × ⁵⁄₁₆" holes, 4 × ¼" holes, or 57 × ¹⁄₁₆" holes (please send pictures if you manage to pull off the latter).

Fluid Mechanics

The same projectile motion method can be used to determine how far the water will travel out of the spout in the large fountain. However, we need to get the water the rest of the way down the fountain without splashing out, and for that, we will have it run down two chains. This works due to a combination of forces, namely: the water molecules' tendency to stick together, known as *cohesion*, and the water molecules' tendency to stick to the solid surface of the chain, known as *adhesion*.

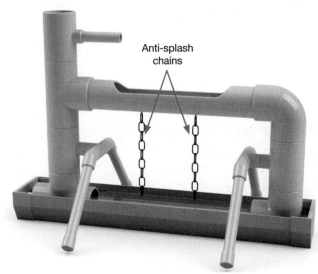

Anti-splash chains

The combination of adhesive and cohesive forces for a particular solid/liquid pairing can be described by measuring how well the liquid wets the solid. Yes, there are different degrees to which a solid can get wet. If the cohesive forces are much stronger than the adhesive forces, meaning the liquid stays together really well but is repelled by the solid, then it will have low *wetting*. On a freshly waxed car, water tends to bead up and roll off—an example of poor wetting (but good automotive care). On the other hand, if the cohesive forces are much weaker than the adhesive forces, then the solid will have high wetting. If you haven't waxed your car in a while, you might notice that when it gets wet the water doesn't bead up; it instead spreads out and dries, leaving streaks that expose your lazy habits.

Low wetting ⟶ High wetting

The chains on the fountain keep the water close to the surface of the links, preventing the flow from separating, which would result in the wetting of your floor. The cohesive forces keep the water together so that as it transitions down the chain, any water that is separated from the chain stays tied to the rest of the flow. This may take some adjustments with the chain attachments and flow rate, but if balanced well, the flowing water will form a beautifully smooth membrane across the links in the chain until your thirsty cat laps it up.

Engineering Application

This concept of wetting really does have many other applications in engineering. One example is called *liquid penetrant inspection*. It's a simple test used for identifying small cracks in structures. First, a liquid fluorescent dye is applied to the surface of the structure. The dye will penetrate any cracks and the excess is wiped away. When inspected with an ultraviolet light, the dye that has seeped into the cracks will light up, highlighting the damage. If the liquid dye and solid structure have low wettability, the dye will bead up and will not be able to penetrate. So it is important that the dye can adequately wet the surface of the structure. The higher the wettability, the smaller the cracks that can be identified.

Small Drinking Fountain
Tools

#	Description	Potential Alternative
T1	Handsaw	Rotary tool
T2	Drill and drill bits	-
T3	File	Rotary tool, Sandpaper
T4	Scissors	-

Materials

Dimensions are in inches.
Drawings are not to scale.

#	Description	Size	Quantity
P1	PVC coupling*	2"–¾"	2
P2	PVC coupling	2"–2"	1
P3	PVC coupling	¾"–½"	2
P4	PVC pipe	½" by 2' long	1
P5	Plastic tubing	0.625" OD (outer diameter), ½" ID (inner diameter), 6" long	1
P6	Plastic tubing	½" OD, ⅜" ID, 6" long	1
P7	Desktop fountain pump	Small enough to fit under the 2"–2" coupling, ≈ 40 gallon per hour capacity	1
P8	Timer (optional)	Find one commonly used to set yard lighting displays	1

* Couplings are sometimes labeled as "bushings"; either way, they are in the plumbing section near all the rest of the PVC pipe.

Step 1 Insert a 2"–¾" coupling into the 2"–2" coupling.

Orient the couplings as shown. Cut a 2" slot in the two pieces together, from the bottom up. Drill a small hole (about ¼") at the end of the slot to form a guide for the pump's cord.

Tools: T1, T2
Materials: P1, P2

Step 2 Drill small drainage holes as shown.

Still water is bad. Without these drainage holes, the compartments will fill up, causing water to get trapped, which would quickly result in a funky-smelling fountain.

Tools: T2

Step 3 Modify one of the ¾"–½" couplings.

Inside the coupling is a lip to prevent the ½" pipe from pushing all the way through. File down this lip so that a pipe can be inserted from either end.

Approximate
location of inner lip.

Tools: T3
Materials: P3

Step 4 Connect the modified ¾"–½" coupling with an unmodified one using a piece of ½" PVC pipe about 1" long.

Insert these pieces into the assembly with the modified coupling on top.

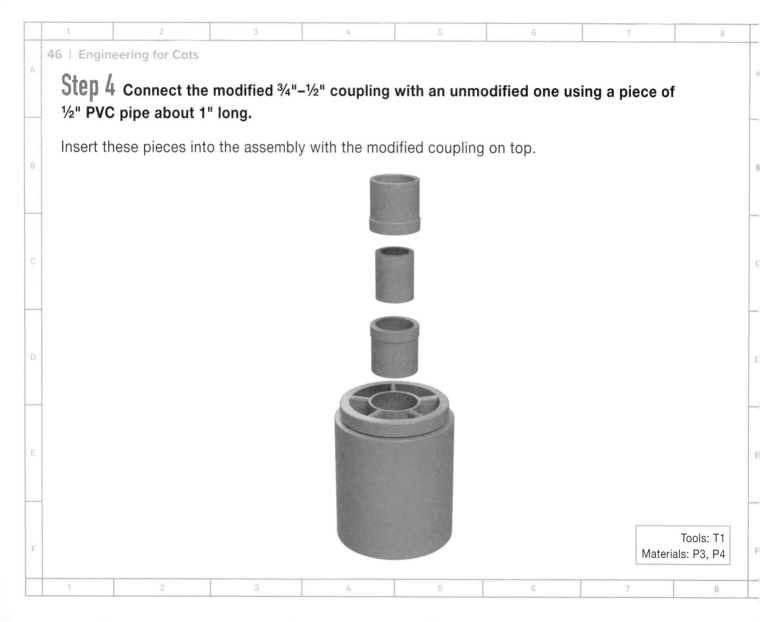

Tools: T1
Materials: P3, P4

Step 5 Cut or drill an exit pattern in the remaining 2"–¾" coupling (the top section) to allow the water to trickle out.

You can skip this step and just have the water overflow from the top, but make sure to drill a small drainage hole so that the water still circulates.

Possible water exit patterns

The water should have no trouble making it over the top even with a small drainage hole added beside the upper opening.

Tools: T1, T2
Materials: P1

Step 6 Fit the top section you created in Step 5 over the ¾"–½" coupling.

Insert a piece of ½" PVC pipe about 3" long into the coupling, which will fit if you modified it correctly in Step 3.

3"

Tools: T1
Materials: P4

Step 7 Insert a piece of the ½" ID (inner diameter) clear tubing up into the ½" PVC piece used to connect the two couplings in Step 4.

It should fit snugly.

The ⅜" ID tubing will fit into this tubing and can be used to connect to the pump.

Tools: T4
Materials: P5, P6, P7

Step 7 (Optional)

If you'd like, you can hook up the pump to a timer commonly used for yard lights.

This timer will put the fountain on a regular schedule, so your cat can get routinely excited about it, but it is also helpful because you don't want to have the fountain running all the time. If the fountain is always on, your cat will be constantly trying to empty it, so there is a risk of it burning out. A timer will ensure that the pump remains submerged in water.

Materials:P8

Complete!

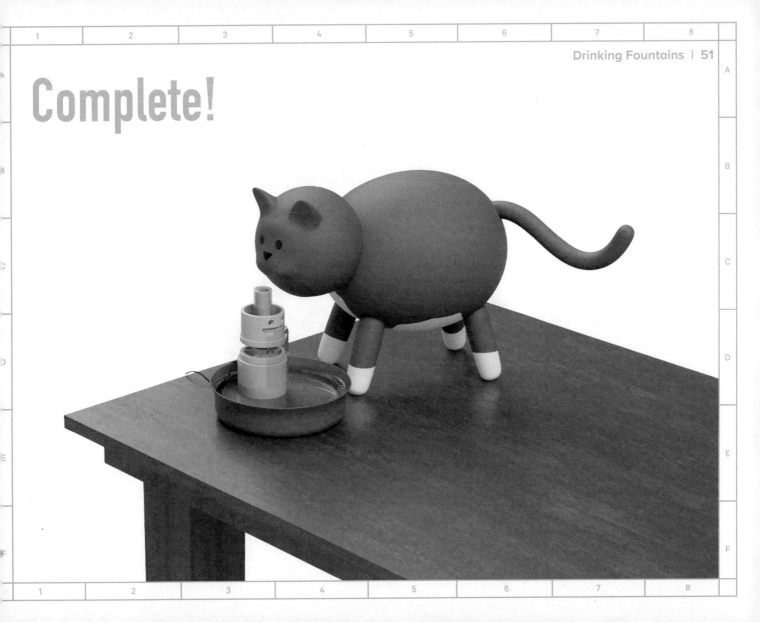

Large Drinking Fountain

This self-contained drinking fountain is designed to maximize the amount of water features without becoming overly complex. It also incorporates a standard water filter.

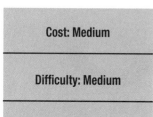

Cost: Medium

Difficulty: Medium

Estimated Build Time:
3 to 4 hours

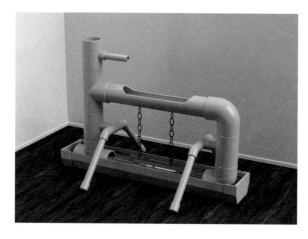

This water feature is so popular with my cats that it has earned the special name "Niagara Paws."

Tools

#	Description	Potential Alternative
T1	Handsaw	Rotary tool
T2	Drill and drill bits	-
T3	Scissors	-

Materials

#	Description	Size	Quantity
P1	PVC coupling	2"→½" T-coupling	3
P2	PVC elbow joint	2"	2
P3	PVC T-joint	2"	2
P4	PVC pipe	2" by 2' long	1
P5	PVC T-joint	½"	2
P6	PVC 45-degree elbow joint	½"	4
P7	PVC end cap	½"	4
P8	PVC pipe	½" by 4' long	1
P9	Plastic-coated chain	Small links (about 1"), at least 2' in length	1
P10	Tie wraps (zip ties)	Small	2
P11	Plastic tubing	0.625" OD (outer diameter), ½" ID (inner diameter), 6" long	1
P12	Plastic gutter	25" long, large enough to accommodate 2" PVC pipe	1
P13	Plastic gutter end caps	Match gutter size	2
P14	Desktop fountain pump	Small enough to fit a 2" T-piece, ≈60 gallon per hour capacity	1
P15	Water filter	2" diameter filter (The same kind that are used for standard pitchers.)	1

Dimensions are in inches. Drawings are not to scale.

Step 1 Cut a piece of 2" PVC pipe approximately 15" long, cut out a section on the top, and drill four holes in it, as shown.

This piece will be the top water receptacle. The cutout should go a little less than halfway through the pipe diameter. It is easiest to cut with a rotary tool, but a fine-tooth handsaw can work as well.

Drill two pairs of small holes through the bottom of the receptacle, large enough to accommodate tie wraps. The tie wraps will be threaded through to attach the two chains.

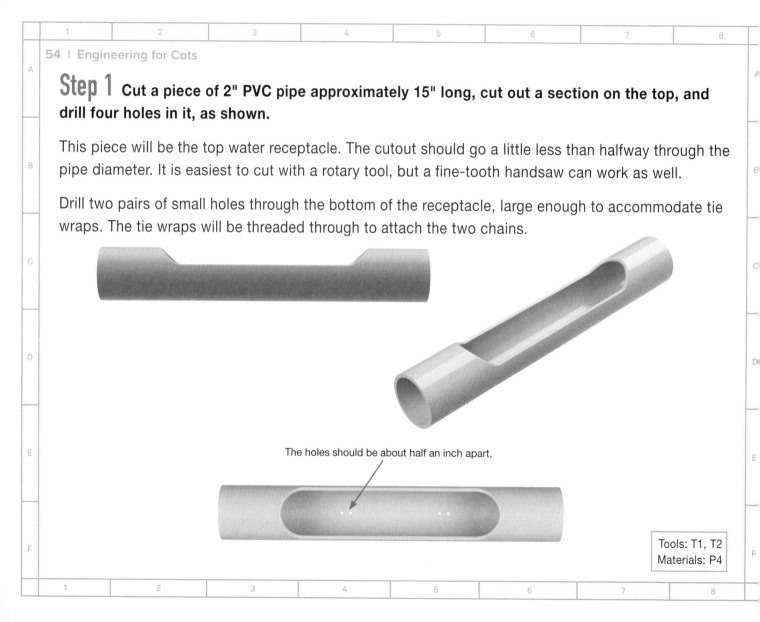

The holes should be about half an inch apart.

Tools: T1, T2
Materials: P4

Step 2 Add joints to the water receptacle.

Attach a T-joint to one side of the receptacle. Attach an elbow joint to the other.

Materials: P2, P3

Step 3 Cut two pieces of 2" PVC pipe approximately 1½" long.

Use these as connectors to attach two 2"→½" T-couplings as shown.

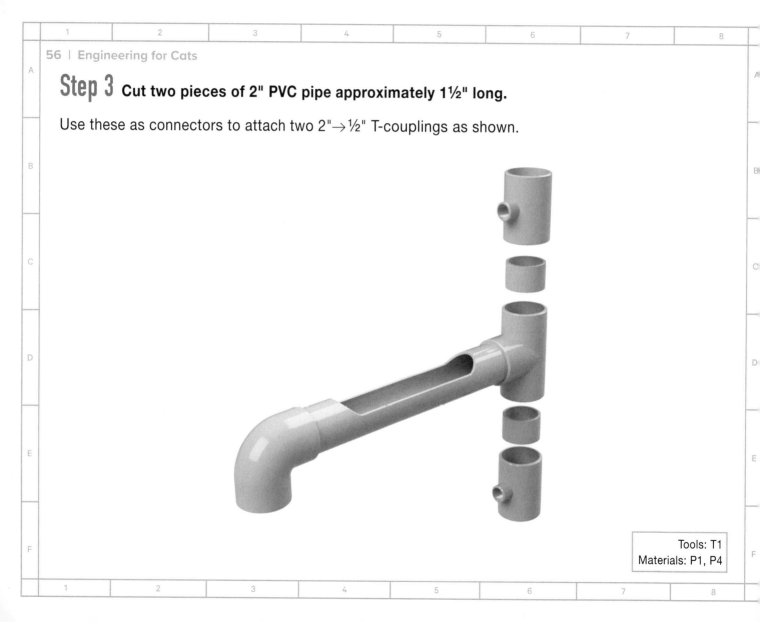

Tools: T1
Materials: P1, P4

Step 4 Cut another 1½" length of 2" PVC to attach the same type of coupling on the other side.

If you would like to include a water filter, a standard pitcher-type filter (2" diameter) will fit right into the PVC connector you cut in Step 3. Not all of the water will flow through the filter on every cycle because it can go down the chains or the other leg, but enough will go through it to make a difference.

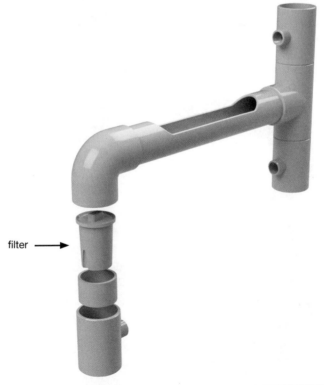

filter ⟶

Tools: T1
Materials: P1, P4, P15

Step 5 Cut two more 1½" connector pieces from 2" PVC to attach another T-joint and elbow joint as shown.

Also cut a short piece of ½" pipe and attach to the top adapter as the spout.

Tools: T1
Materials: P2, P3, P4, P8

Step 6 Construct the legs.

Fit together one T-joint, two 45-degree elbow joints, two 7¼" lengths of PVC, and two end caps. All of these parts will be ½" PVC.

MAKE 2

T-joint

1" length PVC

45-degee elbow

7¼" length PVC

end cap

Tools: T1
Materials: P5, P6, P7, P8

Step 7 Attach the two sets of legs to the fountain.

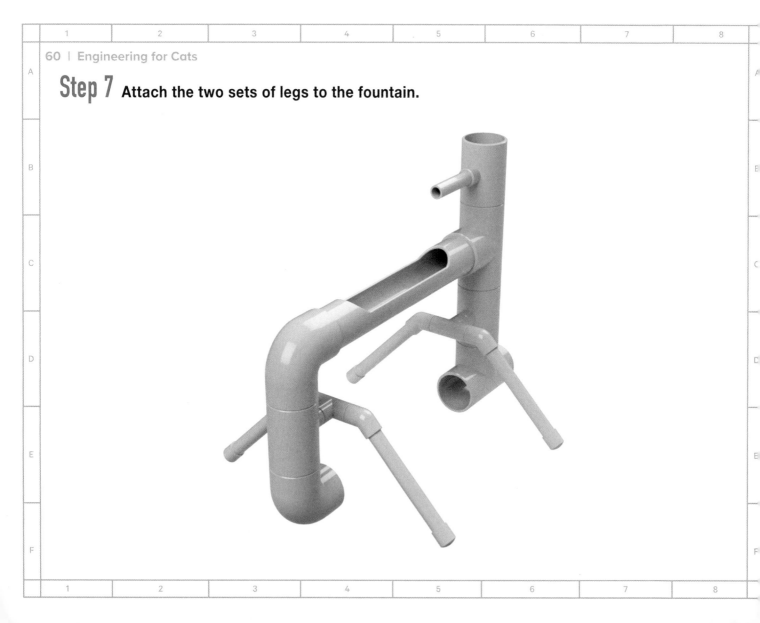

Step 8 Separate the chain into two lengths.

Measure the length you'll need for each chain. Use the saw to cut one link out so you have two chains.

Loop a tie wrap through each pair of holes in the top water receptacle to attach the chains so they hang as shown below. You do not need to pull the tie wraps tight, as some slack may help the water transition to the chain. This transition may take some fine-tuning—adjust—the size of the holes and how tight the tie wraps are pulled. When done successfully, the water should flow down the chains and into the bottom receptacle without any splashing.

Tools: T1
Materials: P9, P10

Step 9 Place the pump in the bottom T-piece and run the plastic tubing up to the ½" PVC spout.

Tools: T3
Materials: P11, P14

Step 10 Cut a section of plastic gutter to the appropriate length and install the end caps.

Tools: T1
Materials: P12, P13

Complete!

THE CAT CAVE

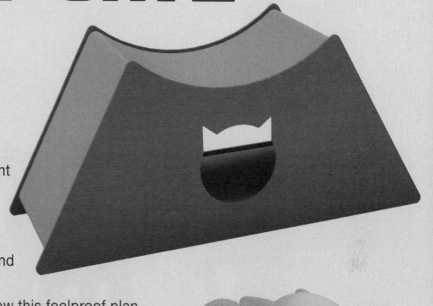

Problem

Cats enjoy destroying your upholstered belongings.

Protecting the Human Furniture

Damaged furniture is often the tipping point for some people when deciding whether to get a cat. They might like cats just fine, but not enough to justify having their furniture torn up or covered in fur. If you find yourself with a roommate like this and it is preventing you from having a cat, just follow this foolproof plan and you should have no further issues in adopting a fuzzy friend:

Get a new roommate. Your current one clearly exhibits poor decision-making skills.

If you are unable to do this for whatever reason, then you can move on to . . .

Solution

1. Add some cat furniture to your living room so the human furniture won't be torn up.

2. Make sure the cat furniture is awesome. In this case, a dual-function cat scratcher and lounge chair made from plywood.

"But how will I be sure the cat will scratch the cat furniture and not the human furniture?" you ask. First of all, doesn't your cat have a name? He might be scratching up your stuff so much because you show him disrespect by referring to him as "the cat."

Second, refer to Step 2: *Make sure the cat furniture is awesome.* You have to provide furniture that "the cat" will *prefer* to nap on and scratch. The choice of what to scratch will always include your sofa, so the goal is to provide a more favorable alternative.

Cost: Medium

Difficulty: Medium

Estimated Build Time:
6 to 8 hours

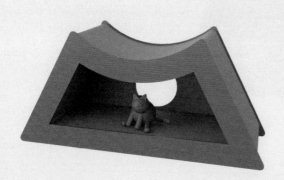

Cats seem to choose what to scratch partially out of habit—
scratching the same spot every day can be part of a routine
that makes them feel at home. If you are building this cat
cave to help protect an already-scratched sofa, then it might
take a little encouragement for your cat to make the switch.
Temporarily putting tape over the spots on the couch that are
often scratched can help him break the habit. If that doesn't
work, then I recommend lounging in the cat cave yourself until
your cat gets jealous and switches over just to spite you.
(OK, filling it with delicious treats might work, too.)

FEATURES OF THE CAT CAVE

- Angled surfaces for optimal scratching

- Curved top for ergonomic lounging

- Carpeted interior for protected napping

- Escape route out back for emergency exits

Bonus: Can serve as a nice bunker where a cat can go to feel protected without being isolated in a closet or under a bed.

ENGINEERING 101
Composite Materials

Over the last several decades, composite materials such as carbon fiber have seen a rapid growth of applications. These materials, made up of layers of fabric infused with a liquid polymer (basically plastic), can be stronger than steel while remaining incredibly lightweight.

These can be expensive to manufacture and are really used only where weight is a strong design requirement, such as in rockets or airplanes. Although this cat cave does not directly use these advanced composite materials (yet), some concepts of the material science are applicable in two important steps. So after constructing this project, make sure that your new résumé includes composite material manufacturing.

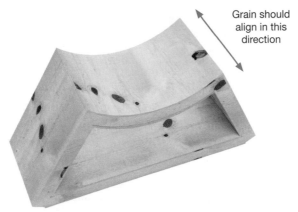

Grain should align in this direction

Fiber Direction

The strength of carbon fiber reinforced polymers (or CRFPs) is not consistent in every direction. The material is strongest in the direction in which the fibers run, but if loaded in other directions it mainly relies on the polymer material for strength. This is why several layers of fabric are usually stacked on top of each other, with the fibers of each layer in different directions. This same principle applies to the nonadvanced composite material that is used in the cat cave: plywood. Plywood consists of several layers of wood with the grain running in different directions. Taking advantage of the grain direction will make construction much easier.

The ¼" plywood used on the frame of the cat cave will likely be made up of three layers of wood. For the flat pieces on the sides and bottom of the cave, it doesn't really matter which way the grain is oriented. For the two pieces that cover the top, however, the grain will need to be oriented correctly in order for it to easily form to the curvature of the cave.

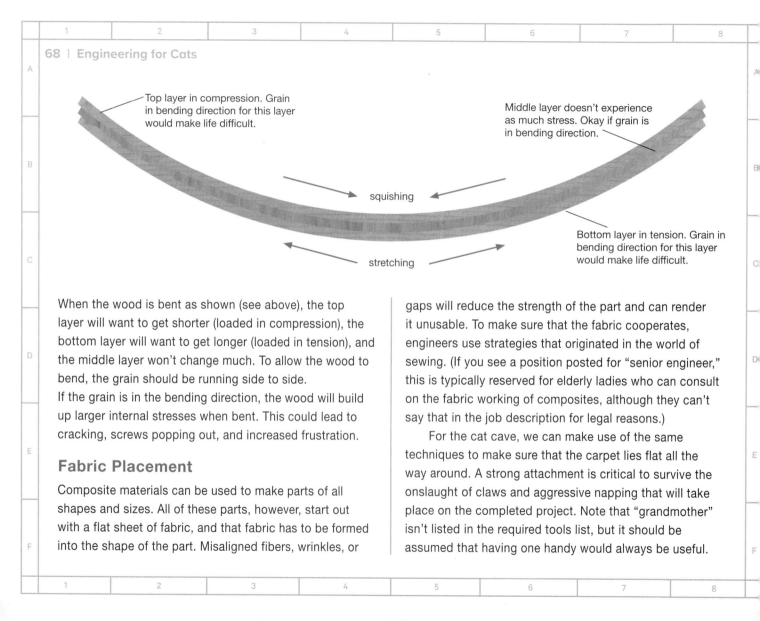

Top layer in compression. Grain in bending direction for this layer would make life difficult.

Middle layer doesn't experience as much stress. Okay if grain is in bending direction.

squishing

Bottom layer in tension. Grain in bending direction for this layer would make life difficult.

stretching

When the wood is bent as shown (see above), the top layer will want to get shorter (loaded in compression), the bottom layer will want to get longer (loaded in tension), and the middle layer won't change much. To allow the wood to bend, the grain should be running side to side.

If the grain is in the bending direction, the wood will build up larger internal stresses when bent. This could lead to cracking, screws popping out, and increased frustration.

Fabric Placement

Composite materials can be used to make parts of all shapes and sizes. All of these parts, however, start out with a flat sheet of fabric, and that fabric has to be formed into the shape of the part. Misaligned fibers, wrinkles, or gaps will reduce the strength of the part and can render it unusable. To make sure that the fabric cooperates, engineers use strategies that originated in the world of sewing. (If you see a position posted for "senior engineer," this is typically reserved for elderly ladies who can consult on the fabric working of composites, although they can't say that in the job description for legal reasons.)

For the cat cave, we can make use of the same techniques to make sure that the carpet lies flat all the way around. A strong attachment is critical to survive the onslaught of claws and aggressive napping that will take place on the completed project. Note that "grandmother" isn't listed in the required tools list, but it should be assumed that having one handy would always be useful.

Technique 1

Placing carpet around an inside curve

When folding around an inside curve, the material will want to pull apart; you need to make a few straight cuts in the material hanging off the edge, so that it will lay flat. The industry term for these cuts is *darts*.

Technique 2

Placing carpet around an outside curve

For an outside curve, the material will want to fold up on itself. To allow it to lie flat, cut out a few triangles in the material hanging off the edge. These would be referred to as *V-darts*.

Technique 3

Turning the carpet around an outer edge

To turn a corner, cut out another triangle over the corner, as shown. It's okay if it overlaps a little bit, but you don't want to bulk up the side, because the end piece will be placed on top of these folds (see Step 9 of the Cat Cave instructions).

Tools

#	Description	Potential Alternative
T1	Drill and drill bits	-
T2	Jigsaw	None (this one requires a lot of cutting)
T3	Staple gun	Hammer and tacks
T4	Scissors	Box cutter
T5	Finishing materials (sandpaper, and nontoxic paint or stain, as desired)	-

Materials

#	Description	Size	Quantity
P1	¼" plywood	4' × 8'	1
P2	⅜" plywood	4' × 4'	1
P3	2 × 4 stud	8' long	2
P4	Wood screws	1¼" long	1 package (≈100 count)
P5	Staple gun staples	⅜"	1 pack
P6	Carpet	Area rug 5' × 7' or 6' × 8'. Make sure it is durable enough not to be pulled apart by cat claws.	1
P7	Carpet tape	1 roll, 2" wide	1 roll

Dimensions are in inches. Drawings are not to scale.

Step 1 Cut out two ends from the ¼" plywood to support the frame.

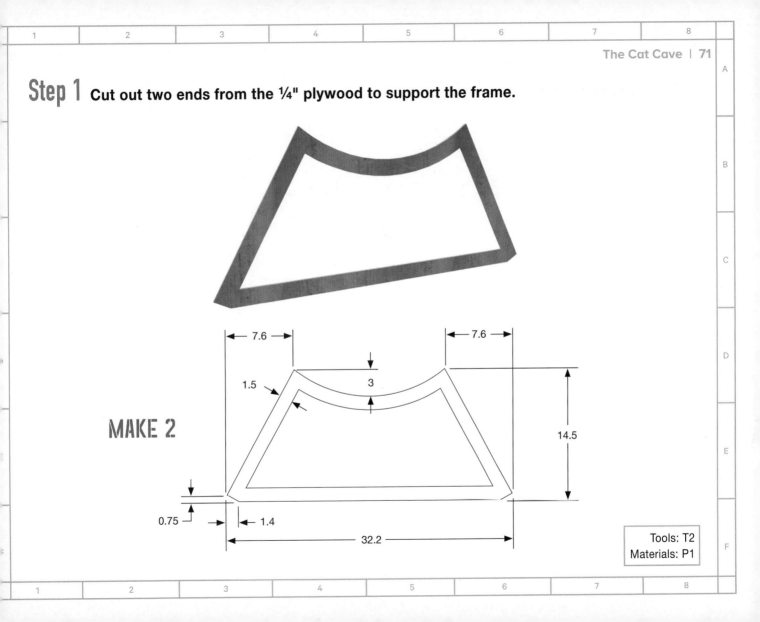

MAKE 2

7.6

7.6

1.5

3

14.5

0.75

1.4

32.2

Tools: T2
Materials: P1

Step 2 Cut twelve support beams out of the 2 × 4 studs for the frame.

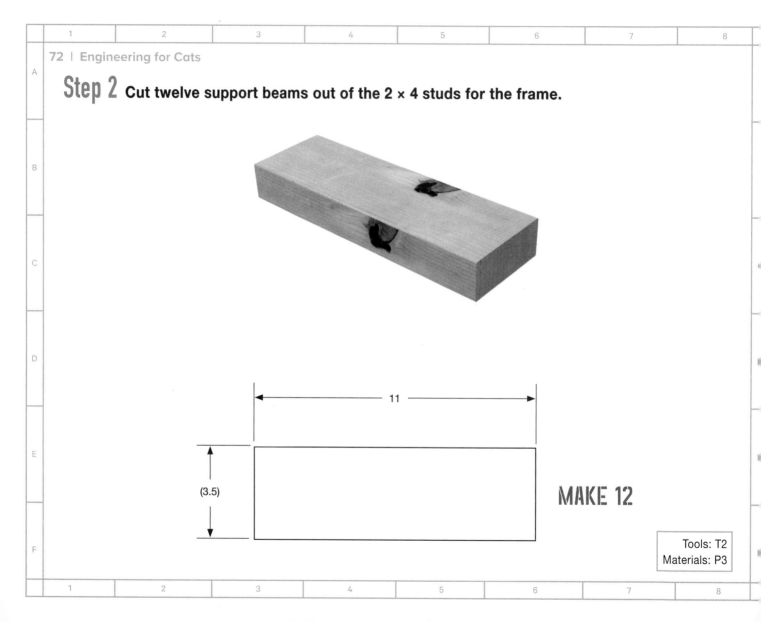

11

(3.5)

MAKE 12

Tools: T2
Materials: P3

Step 3 Screw the beams to the ends of the frame with wood screws, using two screws on each end of each beam.

Locate beams approximately as shown.

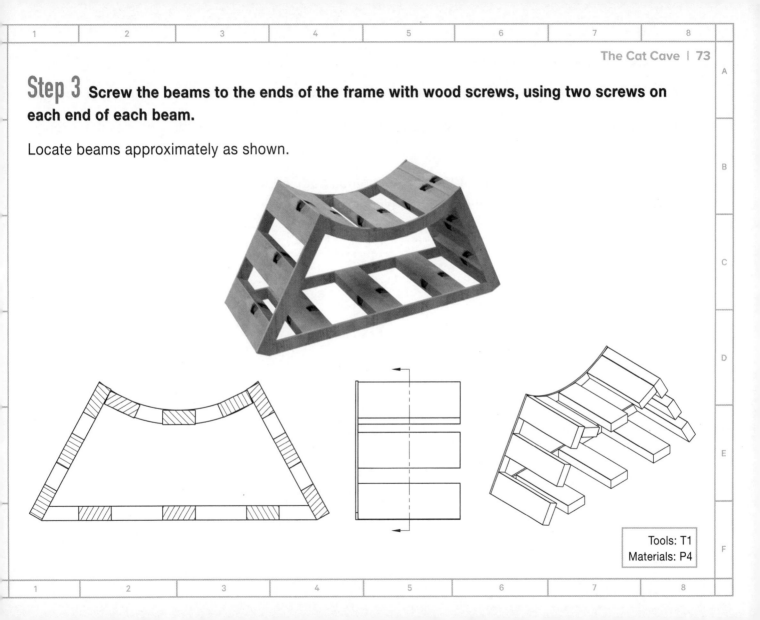

Tools: T1
Materials: P4

Step 4 Measure the frame that you made in Step 1 and cut the ¼" plywood to fit each section.

Dimensions may vary depending on how your frame came out, so use the measurements below only as a guide.

Before you cut, reference the Fiber Direction section on page 67 to make sure the grain direction of the plywood is set correctly. Then cut out the pieces.

MAKE 1 EACH

A — 11.5 × 16.8

B — 11.5 × 19

D — 11.5 × 27.6

F — 11.5 × 29.5

MAKE 2 EACH

C — 11.5 × 11.7

E — 11.5 × 15.7

Tools: T2
Materials: P1

Step 5 Attach the panels to the frame with wood screws.

Do not worry about perfection—these screws will be covered with carpet, so small gaps are acceptable.

Tools: T1
Materials: P4

Step 6 **Cover the frame in carpet.**

Use one piece for the outside and another for the inside. The seam can be hidden on the bottom for the outer piece and in one of the upper corners for the inner piece.

Lay two to three strips of carpet tape on the panels of each side before attaching the carpet. Staples should be needed only on the sides and a few at the seam.

Techniques to get the carpet to lay evenly are detailed in the introduction of this project (detailed cuts needed are not shown).

Tools: T3, T4
Materials: P5, P6, P7

Step 7 Cut out the first end piece of the cave.

Trace the inner and outer edges of the frame onto ⅜" plywood, and offset the line ¼" to ½". This will mean that the end piece will form a small lip on the inner and outer edges of the structure.

Sand and paint the end piece to finish.

Offset from the frame by ¼" to ½".

Tools: T1, T2, T5
Materials: P2

Step 8 Cut out the second end piece, matching the dimensions of the outer edge to the first end piece.

Sand and paint to finish.

Note: The hole I used for the cat cutout was about 6.7" in diameter, and my slightly husky cats (seventeen and twenty pounds) have no issues fitting through this opening. If you have cats of a different size, you might want to adjust the hole.

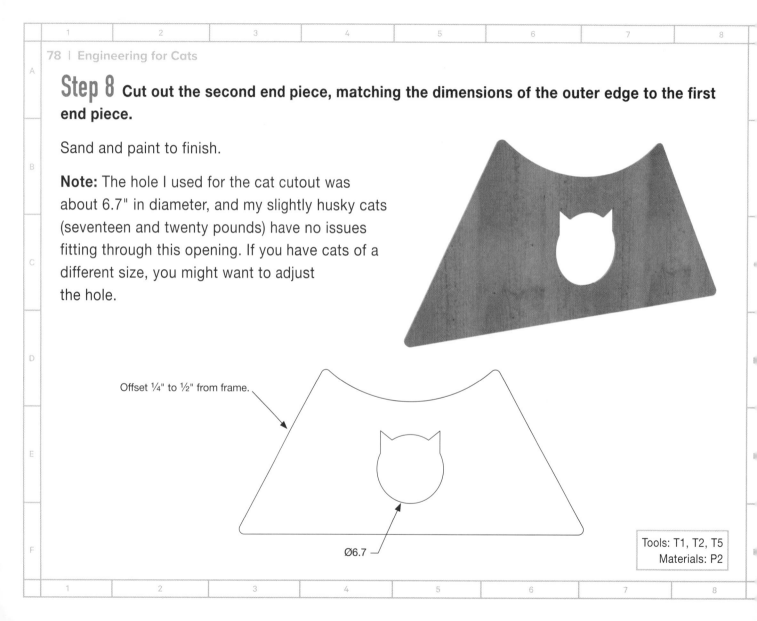

Offset ¼" to ½" from frame.

Ø6.7

Tools: T1, T2, T5
Materials: P2

Step 9 Attach the end pieces to the frame.

To make sure the structure will sit on the end pieces evenly, sit the frame on a scrap piece of plywood before screwing the end pieces on each side.

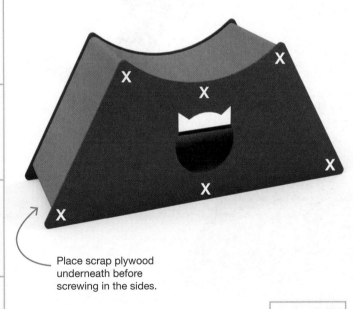

Place scrap plywood underneath before screwing in the sides.

Tools: T1, T5
Materials: P4

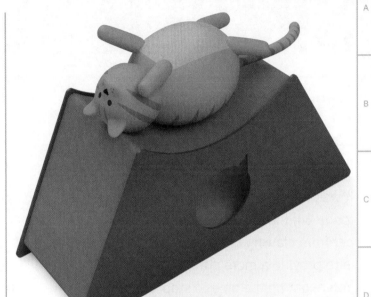

Complete!

SIMPLE SCRATCHER

Problem

Cats need more than one piece of furniture to scratch.

It was mentioned in the Cat Cave project (page 64) that the best way to prevent your cat from clawing up your human furniture is to provide a more attractive alternative. You might not be able to stop your cat from scratching, but you can direct what he scratches. This project is intended to help with that effort in the simplest way possible.

Solution

Provide the cat with one or more carpeted posts designed for scratching.

In many cases, protecting your furniture and legs from cat scratches is simply a matter of providing a series of more appealing alternatives. This simple scratching post will do just that, with the added appeal of being a quick and easy project that usually comes out much prettier than commercially available options.

NOTE: You could make just one of these, but this project is designed to be made from scrap materials. If you are buying materials specifically just to complete this project, consider using the alternative material list and building a handful of scratching posts. Then you can give some to your friends, donate a few to an animal shelter, or just distribute them all over your house to really let your cats feel like royalty.

Cost: See note	
Difficulty: Low	
Estimated Build Time: 1 to 2 hours	

THE SIMPLE SCRATCHING SOLUTION

This is a very easy build to complete, and these types of scratching posts can be placed anywhere, which is perfect for targeting problem areas. If your cat regularly attacks a section of carpet or the leg of a chair and you want him to stop, placing one of these nearby should redirect his attention right away. In some cases it might take some persuasion via catnip or a spray bottle, but it should not take too much effort to get him to make the switch.

The main issue with building your own scratching post is the cost. You should be able to get a similar product in a store for around $10 to $20. In order to buy all the materials needed, this project will run an estimated $36. So why even propose building this scratching post if it is possible just to buy a cheaper one? Here are the three top reasons:

1. This project is designed to use scrap material that would be commonly available in most garages, especially if any of the other projects in this book had recently been made in that garage. If any of the materials needed for this project are already on hand, the actual cost of building it should be much less. This scratching post is a great way to put scrap material to good use, rather than just throwing it out or having it sit around for years.

2. If you do buy the materials new, the stock sizes and quantities for the materials will allow for multiple scratching posts to be built. The instructions include an extra material list sufficient for building six of them, which will have a cost per unit of around $13. Obviously six scratching posts could go to good use in any home, but if that sounds like too many then you can give the extras as gifts for friends or donate them to your local animal rescue. Scratching posts wear out regularly, so any shelter with cats should be happy to have them.

3. I know I said before that I am no expert in aesthetics, but the tan shag carpet that comes on almost every store-bought scratching post is atrocious. That is not a highly used adjective in my vocabulary, but you have to admit that it is appropriate here. With the thousands of carpet patterns available it is sad that almost every cat tree and scratching post is made to look the same. By making one yourself, you have the ability to choose something that you like.

ENGINEERING 101

New Product Lines Should Be Approached with Caution

Say you are in the market for a new car. You're trying to decide between the latest release of a model that has been in production for years and a brand-new model with all of its new technology. To help make your decision, it may be helpful to consider the production process.

A major influence on the cost of a product is the total quantity that is being produced. To manufacture a single specialized bolt can cost thousands of dollars, because the machinery needed to make a new bolt is expensive to operate, and getting the part built correctly will require costly inspections from humans who usually insist on being paid in money. However, you can easily buy a common bolt in the store for just a few cents, because the factories that produce common bolts are set up to make huge quantities of them. They can invest in creating very reliable and efficient processes using huge and expensive machines, and the large amount of units produced will make up these development costs.

Common bolt: $0.15

Special bolt: $2,000

You can see this in action with all types of products. Two different items might contain similar technology and materials, such as a speaker and a microphone, but they will not cost the same amount to buy (or to produce). Because the market for speakers is far bigger than the market for microphones, the speakers will be much cheaper.

In addition to cost, the quality of a part can increase through large production. After the initial development is completed, a manufacturer can refine the process to continue to save money, eliminate defects, and increase efficiency. This is where we get back to our car-buying question. Auto manufacturers will usually come out with a new version of each of their models every year. Most of the time this involves very minor aesthetic and performance changes, which is to the advantage of both you and the manufacturer. The parts that are working well can continue to remain in production from the previous year's model, and the parts that are causing any issues can be upgraded. This means that the parts inside the car are not only reliable, but more readily available for replacement because most of the parts will be featured on many different models.

If you purchase a car that just entered production, many more of its parts will be in mass production for the first time. Some defects associated with them will not have been identified yet, and many of the parts will have plenty of room for improvement. This is why it usually takes some pretty special creativity and innovation, such as using electric power, to break into the auto industry.

In the case of our scratching post, this has really all been an elaborate excuse for why I failed to design a homemade scratching post that is cheaper than the store-bought option. Making six scratching posts will reduce the cost of the scratching post by more than 60 percent. The fact that producing six already lowers the cost below the store-bought options (when those manufacturers are making tens of thousands) might highlight just how much the stores are marking up their products.

Tools

#	Description	Potential Alternative
T1	Circular saw	Jigsaw, Handsaw
T2	Drill and Drill bits	-
T3	Scissors	Box cutter
T4	Staple gun and staples	Hammer and tacks
T5	Wrench	-
T6	Finishing materials (sandpaper, and paint or stain, if desired)	-

Materials

#	Description	Size	Quantity
P1	4 × 4 Beam	8' piece	1
P2	½" Plywood	2' × 2' (particle board or OSB)	1
P3	Short wood screws	Between ½" and ⅞" long	4
P4	Medium wood screws	Between 1¼" and 2" long	3
P5	⅜" Lag bolt	3" long	1
P6	Flat washer	Large enough to fit the lag bolt	1
P7	Thin rope or bungee cord	Approximately ¼" in diameter, 1' long	1
P8	Small tie wraps (zip ties)	Pretty much as small as they get	3
P9	Crafting feathers	Depends on your cat. I go pigeon-size, but if you want to really treat your cat, you can go condor.	2
P10	Carpet	At least 2' × 4'; you can probably get a carpet runner that will work.	1
P11	Carpet tape	2" wide	1 roll

Alternative Materials List

For building multiple units (sufficient for up to six posts):

#	Description	Size	Quantity
P1	4 × 4 beam	8' Piece	1
P2	½" plywood	4' × 8" (particle board or OSB)	1
P3	Short wood screws	Between ½" and ⅞" long	1 package (50)
P4	Medium wood screws	Between 1¼" and 2" long	1 package (50)
P5	⅜ lag bolt	3" long	6
P6	Flat washer	Large enough to fit the ¼-20 bolt	6
P7	Thin rope	Approximately ¼" in diameter, 1' long	1
P8	Small tie wraps (zip ties)	Pretty much as small as they get	20
P9	Crafting feathers	Depends on your cat. I go pigeon-size, but if you want to really treat your cat, you can go condor.	12
P10	Carpet	6' × 8' area rug	1
P11	Carpet tape	2" wide	1 roll

Dimensions are in inches.
Drawings are not to scale.

Step 1 **Measure and mark the tower piece on the 4 × 4 beam and cut it out.**

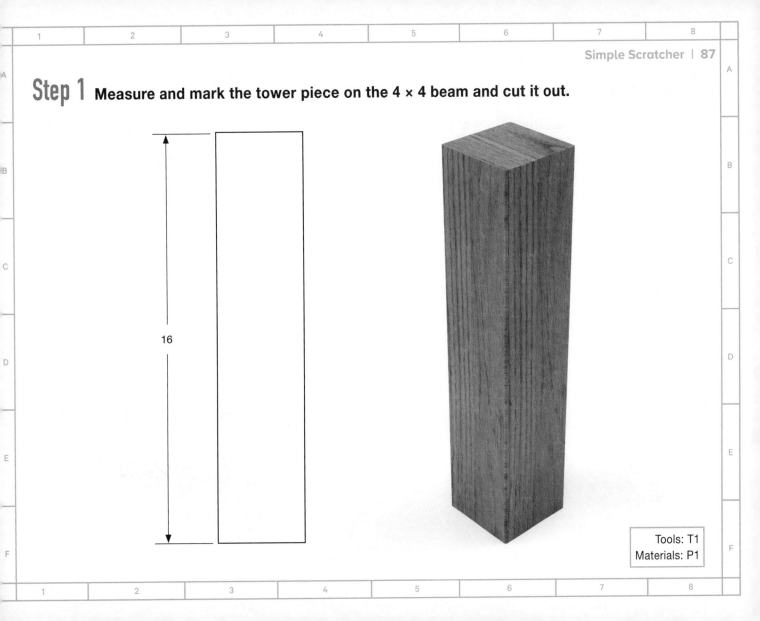

16

Tools: T1
Materials: P1

Step 2 Create the base.

Measure and mark the dimensions of the two squares below on ½" plywood and cut them out.

Check your flat washer to make sure a hole 1¼" in diameter will provide clearance. Then cut or drill a 1¼" diameter hole in the center of one of the base pieces.

Screw the rectangles together with the short wood screws, as shown.

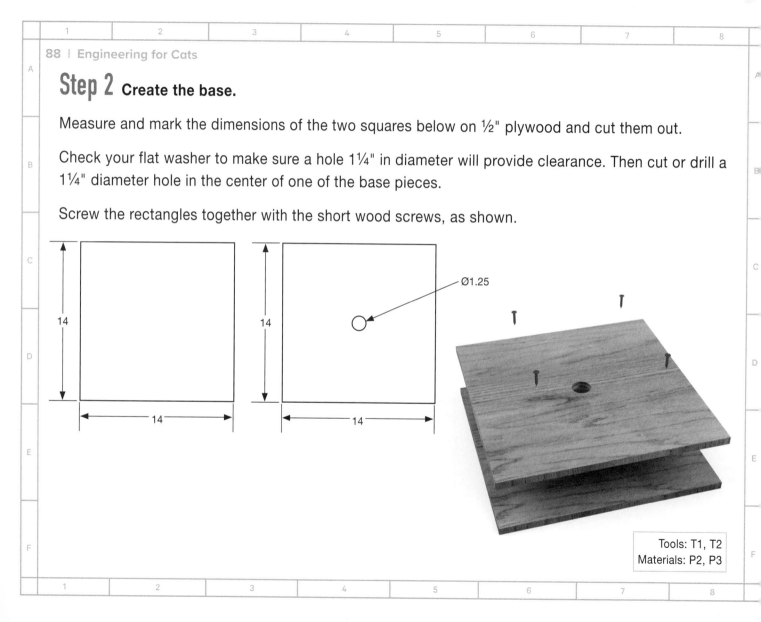

Ø1.25

Tools: T1, T2
Materials: P2, P3

Step 3 Cover the tower and base pieces in carpet as shown. (See page 69 for tips on covering wood with carpet.)

Wrap the tower lengthwise and keep the ends (mostly) free of carpet.

Make sure the side of the base without the hole is completely covered, and the side with the hole is left uncovered. Trim the carpet so that it doesn't double up on itself.

Tools: T3, T4
Materials: P10, P11

Step 4 Cut out the top piece.

Cut a 4¼" × 4¼" square from the ½" plywood.

This piece will not be wrapped in carpet, so spend a little extra time cleaning it up with sandpaper or (optional) painting it.

Tools: T1, T6
Materials: P2

Step 5 **Attach the tower and the base.**

Pre-drill a hole in both the base and tower for the lag bolt. Add the lag bolt and washer, and, for stability, two medium wood screws on either side of the lag bolt.

Tools: T2
Materials: P4, P5, P6

Step 6 Attach the top piece and the rope.

Tie the rope around the screw before tightening the screw all the way.

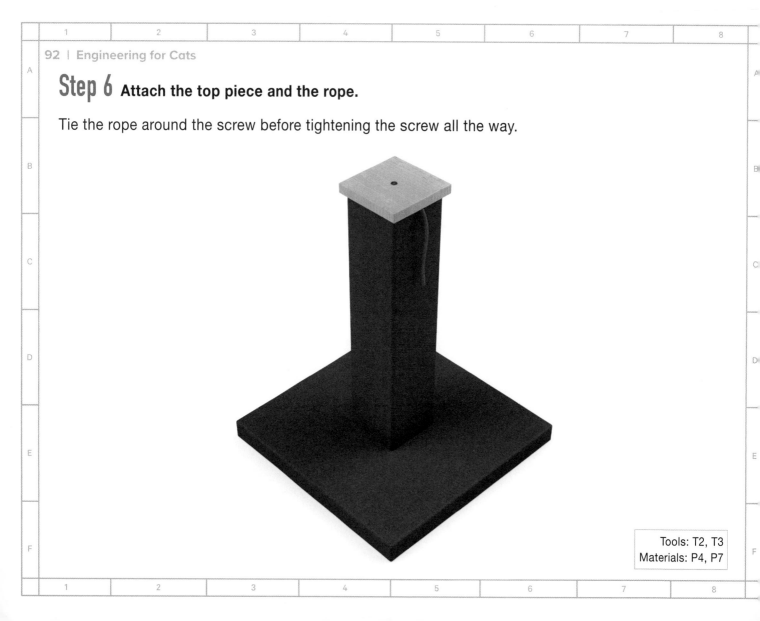

Tools: T2, T3
Materials: P4, P7

Step 7 Attach the feathers.

Use two to three small tie wraps to attach some feathers to the end of the rope.

Tools: T3
Materials: P8, P9

Complete!

BUNK BEDS

Problem

If you have a cat (or an older sibling), you may be familiar with how quickly they will claim your stuff if you let your guard down.

Felines may exhibit this behavior because they don't think anything they own is as important as what you have. Cats are territorial, so when they see where you spend most of your time, such as at your keyboard, on a couch, or in your bed, it makes sense that they would interpret those as places of power—and therefore something they must control.

"You snooze, you lose" seems like an ironic policy for such a sleepy animal to abide by . . .

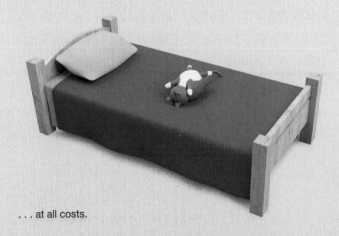

. . . at all costs.

Solution

Bunk beds give cats their own place of power that will interfere less with yours. This model is an easy-to-build and low-cost lounge spot for one or several cats. The top bunk provides a good perch, good for looking out windows and asking for attention. The bottom bunk provides an option for secluded napping without isolating your cat, which can be helpful for getting a more reclusive cat to be a little more social.

You may have to convince your cat that these beds are indeed important, but placing a clean pair of black pants on them or positioning them next to a window should do the trick.

Cost: Medium
Difficulty: Low/Medium
Estimated Build Time: 3 to 5 hours

ENGINEERING 101
Perfection Is Overrated

One of the steps in this project is to cut a length of PVC pipe 9½" inches long. If I sent that instruction to a hundred different factories, I would likely get zero parts that matched up exactly (and that's only a slight exaggeration). The aerospace industry needs to be more specific, so we have an engineer make a model of the part. After two weeks of modeling, six meetings, and twelve redesigns, we get something that looks like this:

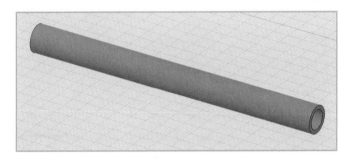

Really showing the skills that come with that expensive engineering degree

This model shows exactly what we had in mind, but building the model to this precision would be a waste of time (and impossible) because we certainly do not need a cut that is 9.500000000000 . . . inches long. Geometric dimensioning and tolerancing (GD&T) is a standard method to show how precise we need this dimension to be. So now we wait another two weeks for the engineer to produce something like this:

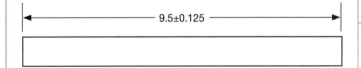

What this image tells us is that if our pipe ends up being anywhere between 9.375 and 9.625 inches long, it will be close enough: 9.5 is the dimension, referred to as *nominal*, and +/- 0.125 is the *tolerance*. The smaller we make the tolerance, the more expensive and time-consuming the part will be to make.

If these bunk beds were made out of a different material, the construction might require tight tolerances due to the number of pieces that need to fit together. The amount that each piece misses the nominal dimension can stack up, and the total error in the assembly might then be enough to transform your project from "functional structure" to "unappealing artwork."

Fortunately, PVC pipe is very forgiving. The tubes are inserted into the joints, which take up the error in the length. For the ½" pipe used for this project, the tubes can be inserted up to ¾" into each fitting, and if you have at least ⅜ of an inch in the joint it should work fine.

That is why for this project all cuts will have a tolerance of +/- 0.125 inches, which should be pretty easy to hit. Of course, the closer the tubes are to the nominal dimension, the easier the project will fit together.

Tools

#	Description	Potential Alternative
T1	Drill and drill bits	Drill press would be helpful
T2	Handsaw	Pipe cutter
T3	Scissors	-
T4	Masking tape	-
T5	Paint	-

Materials

#	Description	Size	Quantity
P1	PVC L-joint	½"	16
P2	PVC T-joint	½"	20
P3	PVC X-joint	½"	4
P4	PVC 45-degree elbow joint	½"	2
P5	PVC pipe	½" × 10' long	3
P6	Flat nylon webbing	1" wide, 25 yards	1

Dimensions are in inches.
Drawings are not to scale.
Tolerance on all cuts is ±0.125 inches unless otherwise specified.

Step 1 Take the 10' PVC pieces and cut them to the necessary lengths.

A handsaw or pipe cutter can work, but for this volume, a miter saw is the most helpful.

New Part

#	Length	Quantity
P5.1	19"	6
P5.2	9½"	6
P5.3	8½"	4
P5.4	4"	4
P5.5	12⅝"	4
P5.6	1"	32

Tools: T2
Materials: P5 × 3

Step 2 In four of the 19" tubes, use a ⁵⁄₁₆" drill bit to drill 18 thru-holes spaced ¹⁵⁄₁₆" apart from each other, and 1½" away from the ends (this may not come out to the exact dimension).

Note that the webbing that threads through them will be 1" wide, but the spacing is slightly less to make the weaving tighter (so paws are less likely to slip through it).

Repeat this step with four of the 9.5" tubes.

11

A

MAKE 4

1.5 ± 0.125   0.9375 ± 0.125

Note: There is no vertical dimension for centering the holes on the pipe. This is partly because it is really difficult to control a hole being drilled into a curved surface. The good news is that even if you mess up it won't matter too much. Once the webbing is added, misaligned holes become much less noticeable.

TYPICAL
(meaning all spaces
between the holes
are this dimension)

Tools: T1
Materials: P5.1 × 4. P5.2 × 4

Step 3 Now is a good time to paint the tubes.

It may be quicker if you paint after the bunk beds are assembled, but then you will see unpainted plastic wherever a gap opens up slightly in the joints.

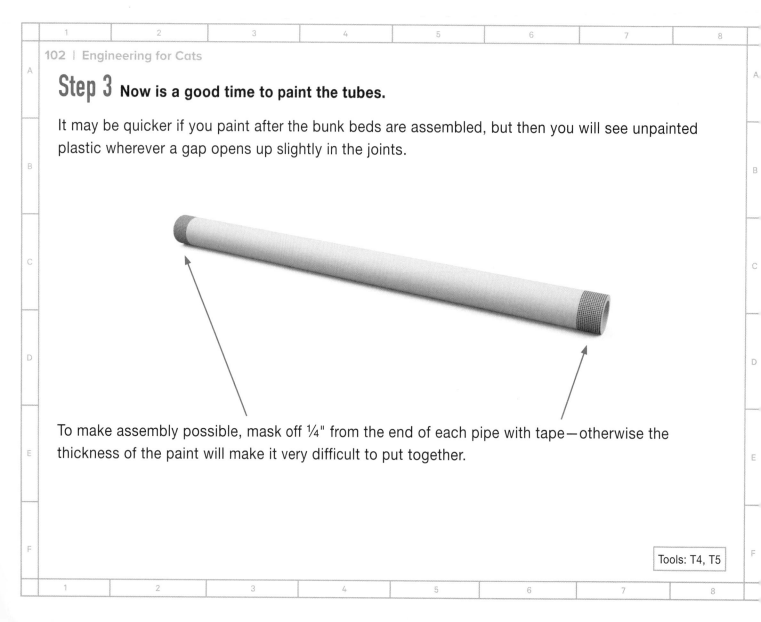

To make assembly possible, mask off ¼" from the end of each pipe with tape—otherwise the thickness of the paint will make it very difficult to put together.

Tools: T4, T5

Step 4 Create the top bunk.

Assemble the top railing of the bunk beds.

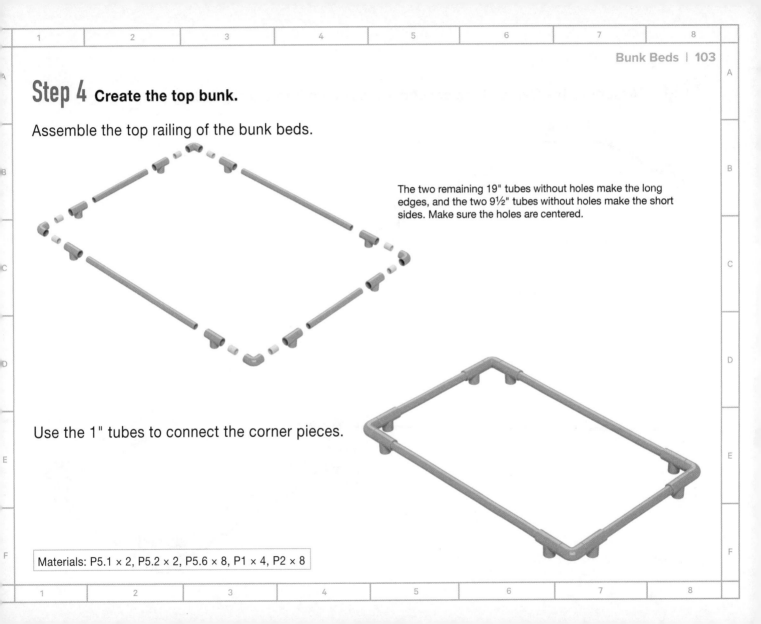

The two remaining 19" tubes without holes make the long edges, and the two 9½" tubes without holes make the short sides. Make sure the holes are centered.

Use the 1" tubes to connect the corner pieces.

Materials: P5.1 × 2, P5.2 × 2, P5.6 × 8, P1 × 4, P2 × 8

Step 5 Assemble the frame of the second level. Attach it to the top railing.

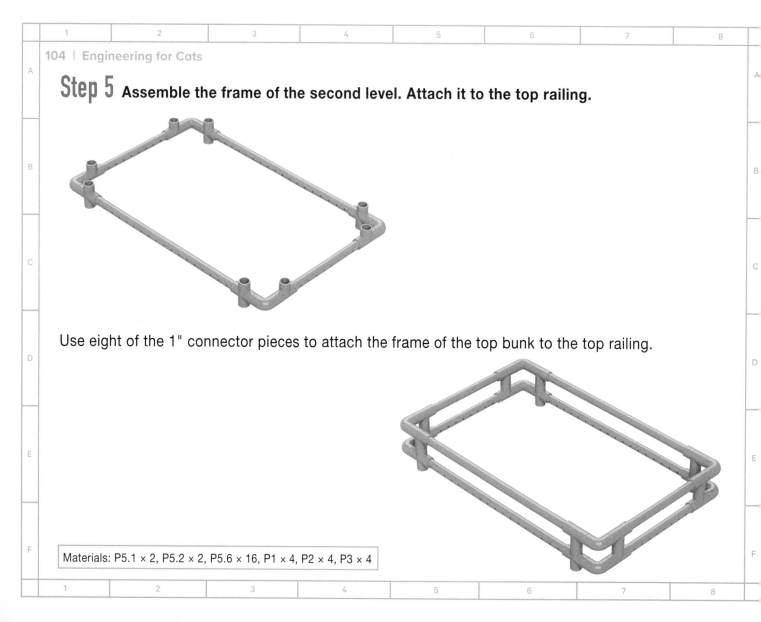

Use eight of the 1" connector pieces to attach the frame of the top bunk to the top railing.

Materials: P5.1 × 2, P5.2 × 2, P5.6 × 16, P1 × 4, P2 × 4, P3 × 4

Step 6 Add the 8.5" tubes to the bottom of the cross connectors.

Materials: P5.3 × 4

Step 7 Assemble and attach the frame of the bottom bunk.

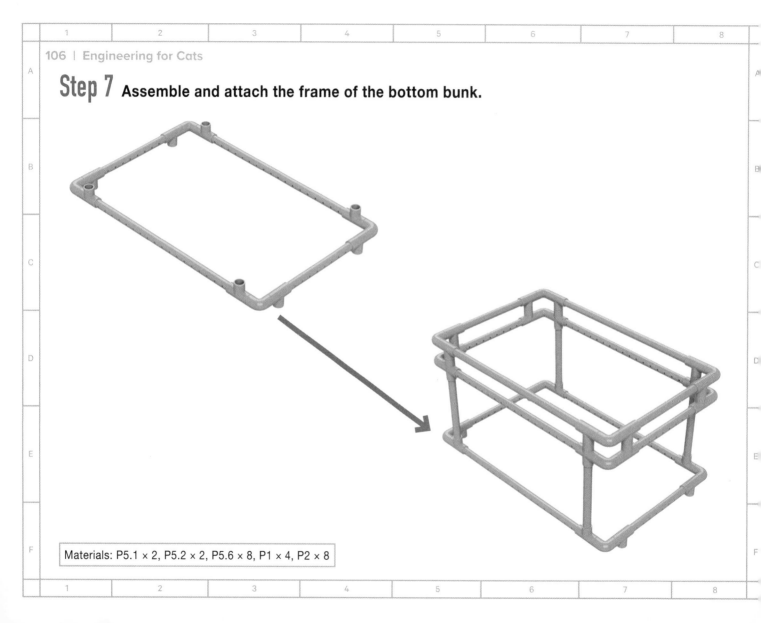

Materials: P5.1 × 2, P5.2 × 2, P5.6 × 8, P1 × 4, P2 × 8

Step 8 Assemble the base.

Use the 4" tubes for the vertical pieces and the 12⅝" pieces for the bottom diagonal pieces.

If the beds are to be placed against a wall, flip around one half of the bottom base to point in so that the structure will be flush with the wall.

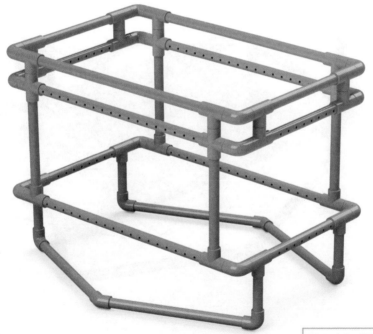

Materials: P5.4 × 4, P5.5 × 4, P1 × 4, P4 × 2

Step 9 Thread the webbing back and forth through the long sides of the bottom bunk and tie off the ends.

If the ends start to fray, you can very briefly burn the edges with a lighter to melt them a bit— although it shouldn't actually catch on fire. When offered as an explanation for why you burned your house down, "I was making bunk beds for my cats" raises more questions than it answers.

Tools: T3
Materials: P6

Step 10 Thread the webbing through the short sides now, weaving over and under the webbing applied in the previous step as you go.

Tools: T3
Materials: P6

Step 11 Repeat Steps 9 and 10 to thread and weave the webbing into the top bunk.

Tools: T3
Materials: P6

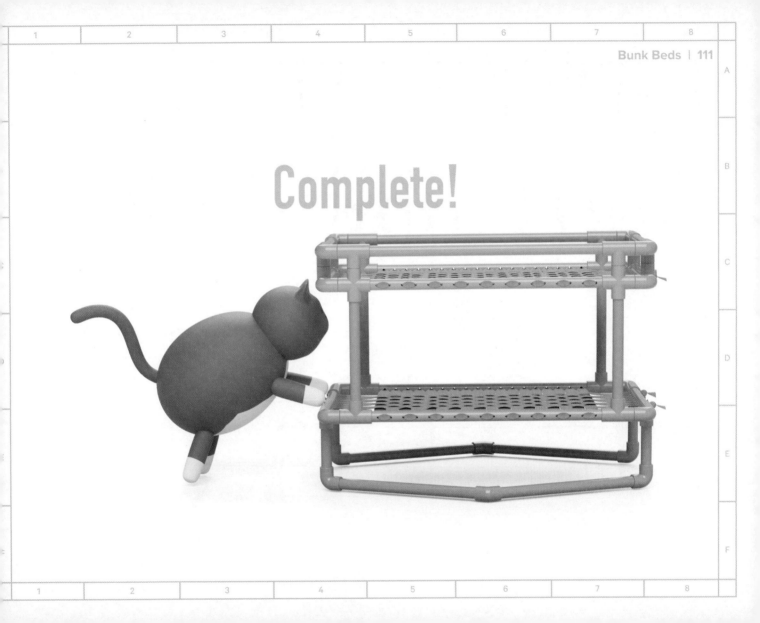

Complete!

CAT TREAT SLOT MACHINE

Problem

Cats will engage in destructive behaviors like knocking things off shelves or terrorizing your other pets when they get bored.

Solution

Provide cats with a bit of entertainment through a treat feeder.

The objective of this device is to mentally stimulate your cat. This is what I have to tell people when they ask if I built this machine solely for my entertainment. Yes, it can be amusing to watch my cat paw at the spinners in the hopes of getting a treat to fall out, all while I kick my feet up, chuckling to myself because I can just eat treats directly from the bag. To some people this seems unkind, so I'm sticking with the other explanation.

With the Cat Treat Slot Machine, treats are fed into the top arm. As your cat spins the device the treats will fall, hitting the pipe in the middle and transferring to one of the other three arms. Only some of the arms have exits, so the treats will fall out only if they end up in that arm as it rotates downward. The first few attempts by your cat might be frustrating to her, because there is no reason your cat should know how to get the treats out. Some light demonstration may be necessary, showing that treats will fall out

Cost:	Low
Difficulty:	Low
Estimated Build Time:	2 to 3 hours

as the arms spin around, but if your cat wants the treats badly enough she will eventually master the operation.

The design presented has three "spinners," each with a different difficulty for getting the treats out. The green is the easiest, because it has two exits for the treats. The yellow also has two exits but is slightly more difficult because it has shorter arms and is in between the other two spinners, making it a little harder to paw at. The red is the difficult one because it has only one exit.

Unfortunately, this toy is effective only if treats motivate your cat. Some cats are just not into treats, and for such cats, you may want to think hard before building this project.

ENGINEERING 101
Probability

This treat feeder is referred to as a slot machine because whether or not a treat comes out on each rotation is based on chance. When the treats fall from the top, there is about an equal chance that they will end up in each of the three lower arms, so the lower arms each have a 33% chance of getting the treat in it and the top arm has a 0% chance of containing a treat.

Consider the red spinner, because it has only one opening. First, if the opening is on top and the treat is added, the treat has a one in three chance of landing in any of the lower arms. The figure below shows each arm labeled with the probability that it contains a treat. Arm 1 is the only one with an exit.

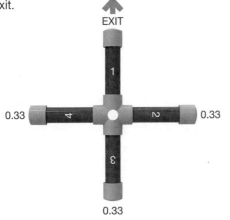

As shown, the device is turned 90 degrees clockwise. If the treat had previously landed in Arms 2 or 3, we can assume it will stay in there, but if it was in Arm 4, which is now at the top, then it will drop again. We don't know if it was in Arm 4; we just know that there was a 1 in 3 chance of that happening. So the chance of the treat being in Arm 1 depends on two events *both* happening:

A: The treat initially landed in Arm 4 (1 in 3)

B: The treat then lands in Arm 1 (1 in 3)

To calculate the probability that two events will BOTH occur, the probabilities of each event happening independently are multiplied.

$$P(A \cap B) = \tfrac{1}{3} \times \tfrac{1}{3} = \tfrac{1}{9}$$

The lowercase *n*-looking symbol above is a symbol called an *intersection*—it indicates that we want to find the probability of both events A and B happening. So there is a 1 in 9 chance that the treat is in Arm 1 after the first turn (0.11). For the treat to end up in Arm 2, only one of the following two events needs to happen:

A: The treat initially landed in Arm 4, and then landed in Arm 2 (1 in 9)

B: The treat initially landed in Arm 2 and then stayed there (1 in 3)

To calculate the probability that *either* of two events occur, the probabilities of each event happening independently are *added*. So in this case:

$$P(A \cup B) = \tfrac{1}{3} + \tfrac{1}{9} = \tfrac{4}{9}$$

The *u*-looking symbol this time is called *union* and indicates that we want to find the probability that either event A or event B occurs. So there is a 4 in 9 chance that the treat is in Arm 2 after the first turn (0.44). The same applies for Arm 3, which also has a 4 in 9 chance of containing the treat after the first turn.

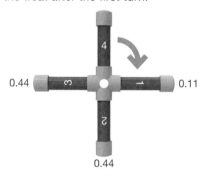

Following this same logic for one more turn shows that there is a 7 in 27 chance that the treat will be in Arm 1

after an additional 90-degree turn. So there is a little better than a 25% chance that the treat will pop out after just two turns. Or if you had put a handful of treats in there, you can expect about a quarter of them to fall out.

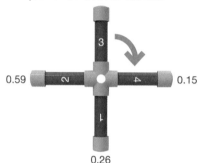

Continuing on with these calculations, we can calculate the odds for each arm after each turn. The next figure shows the next few rotations. Notice that the probabilities no longer add up to 1, because there is a chance the treat would have exited after the second turn.

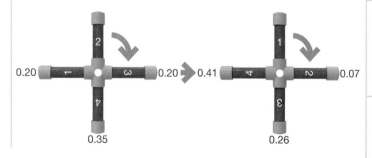

0.40 0.14 0.34 0.13

0.20 0.27

of how long treats might be expected to spin around in each one:

Chance Treat Is Still in Spinner	Number of quarter turns performed	
	Red Spinner	Green/Yellow Spinner
50%	7	2
25%	15	6
10%	23	8
1%	43	16

So with a 26% chance that the treat will fall out after two turns, and a 27% chance that it will fall out after six, altogether that's a 53% chance of getting a treat after six turns. But what if your cat is really unlucky? How long could this treat possibly stay in there? After fifteen quarter turns, there would still be a 25% chance that the treat has not fallen out. After twenty-three, there would still be a 10% chance, and after forty-three there would still be a 1% chance. If you add a handful of treats or dry food to these spinners, then they should continue to fall out after many spins, causing your cat to develop an obsession for batting these spinners around in the hopes of a reward. This obsession is intended to distract from your cat's previous obsessions, which may have included tearing apart your couch or terrorizing the local squirrel population.

Similar calculations can be repeated for the yellow and green spinners as well. The table gives a quick comparison

Probability calculations like this are often used in engineering to avoid doing work. For example, you may have to design a bridge to be able to survive an earthquake and a convoy of tanks driving over it, but if you can show that the chances of both happening at the same time are low enough, you may be able to consider them separately. Then you would not have to design for the combined case of both happening at once.

Tools

#	Description	Potential Alternate
T1	Drill and drill bits (up to 1" bit)	-
T2	Saw	Pipe cutter
T3	Scissors	Box cutter

Materials

#	Description	Size	Quantity
P1	½" PVC pipe	½" × 10'	1
P2	½" PVC T-piece	½"	2
P3	½" PVC L-piece	½"	6
P4	½ → ½" PVC coupling	½"	2
P5	1" PVC pipe	1" × 10'	1
P6	1" PVC X-piece	1"	3
P7	1" PVC end cap	1"	12
P8	Electrical tape	For adding color to spinners (you can use nontoxic paint instead)	3
P9	Poster board	Enough for a 15" × 14" section	1

Dimensions are in inches. Drawings are not to scale.

Step 1 Cut lengths of pipe for the spinners.

Using the 1" PVC pipe, cut the lengths that will be used to make the spinners. The exact length is not too critical—you can even mix and match lengths on each spinner if you want to give your cat an extra challenge. Longer arms make it slightly easier to rotate the spinners.

MAKE 8

5

MAKE 4

4

Tools: T2
Materials: P5

Step 2 Drill holes in the end caps.

Drill holes in five of the twelve total end caps, sanding away any loose pieces. The green and yellow spinners will each get two, and the red spinner will get one. Make the hole big enough so that the treats your cat prefers can fit through it.

Tools: T1
Materials: P7

Step 3 Drill holes in the X-pieces.

This will be a hole just large enough for the ½" PVC pipe to fit through.

Tools: T1
Materials: P6

Step 4 **Assemble the spinners.**

Because the spinners will be dispensing food for your cat, it is a good idea to clean up all the cuts and holes with sandpaper, and thoroughly wash the spinners before moving on to the next step.

Put the two exits next to each other on the spinner instead of on opposite ends of the structure.

EXIT

EXIT

Materials: P5, P6, P7

Step 5 **Wrap the arms of the spinners in colorful electrical tape.**

You could also paint them with nontoxic paint, but that requires waiting for it to dry. Kudos if you have the patience for that.

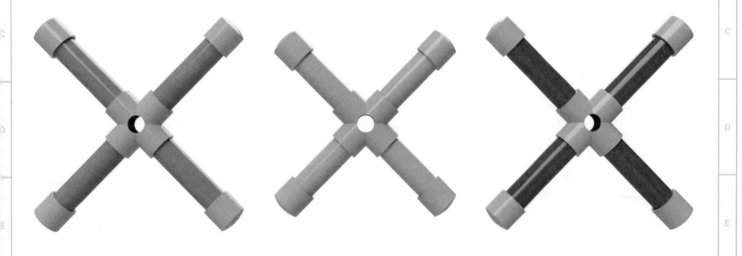

Tools: T3
Materials: P8

Step 6 Using the ½" PVC pipe, cut the lengths of pipe for the base.

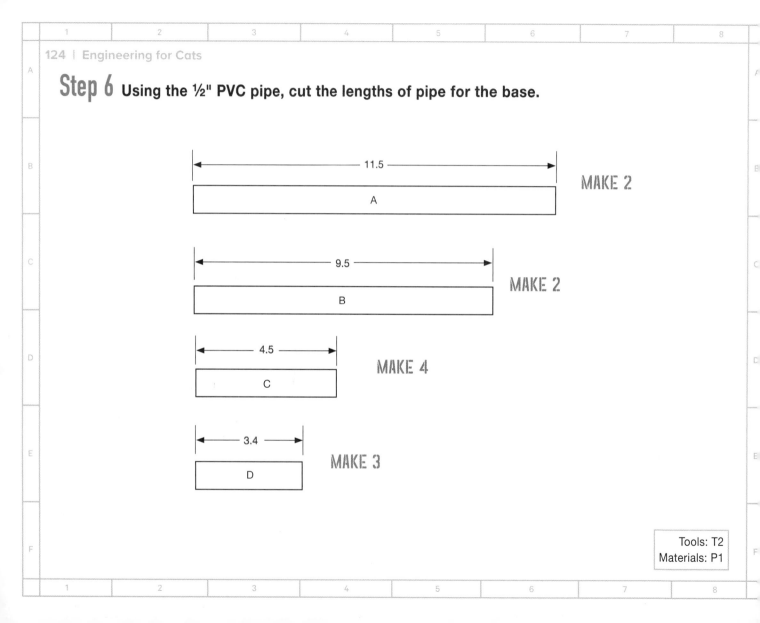

11.5

A

MAKE 2

9.5

B

MAKE 2

4.5

C

MAKE 4

3.4

D

MAKE 3

Tools: T2
Materials: P1

Step 7 Assemble the base, as shown.

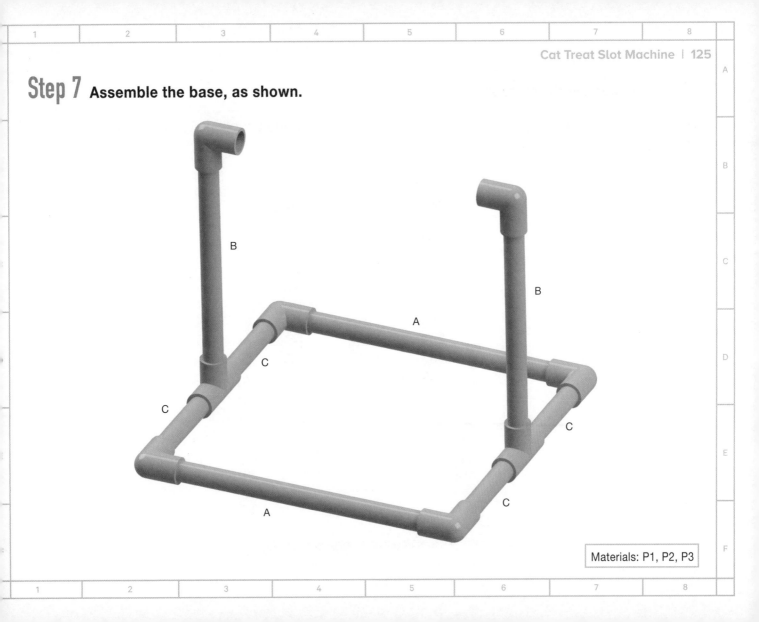

Materials: P1, P2, P3

Step 8 Assemble the spinners.

Use the 3.4" lengths of pipe from Step 6 (length D) along with the couplings to mount the spinners as shown.

Materials: P1, P4

Step 9 Measure, mark, and cut the treat catcher base.

Place it under the spinners.

Tools: T3
Materials: P9

Complete!

DRAWBRIDGE CAT DOOR

Problem

Cats may become restless without special access to the outdoors.

Your standard house usually comes equipped with doors. These are very useful for keeping out terrifying things like bears or for avoiding social contact with other humans, but they can also keep out cats. A thoughtful cat might start to worry about his human if locked out of the room for too long, and, in an effort to comfort his human companions, may resort to singing reassuring melodies outside the door.

All cats know that humans need the most reassurance in the wee hours of the morning.

Solution

Construct an elaborate special access door just for cats.

The Drawbridge Cat Door is a blatant overcomplication of the existing pet door flap design. Its tactical advantages in no way justify the extra effort required to construct it, and it could therefore rightfully be called impractical. But it was not designed for the practical. It was designed for the brave. It was designed for those who refuse to ask if something *should* be built until it has already gained self-awareness and declared war on all of humanity.

It is consequently for anyone who wants to provide special access through a door for their cats, but only if they want that access way to be *magnificent*.

Cost:
Medium

Difficulty:
High

Estimated
Build Time:
6 to 8 hours

The style of cat door presented in this section is based on the drawbridge. Made famous by medieval castles, the unique and elegant drawbridge-style door has more engineering applications than any other door short of the Stargate.

"But why make a drawbridge-style door? Wouldn't it be much cheaper, easier to build, and easier for a cat to use if it hinged from the top?" asked the hater.

The type of door this hater is describing is not a door at all—it's a flap. You probably don't need any help to design a flap. For those who want to give their cats the elegant entryway they deserve, I recommend stepping it up to the same kind of door used to welcome home knights as they returned from slaying dragons.

This particular door is made for interior rooms, and won't seal up well enough to be used for outdoor access. The debate on whether cats should be allowed outdoors or be kept inside is still open. (I personally keep my cats inside because the area in which I live has an abundant coyote population, and their hunting techniques are improving by the day. In most areas, outdoor cats do just fine, but that needs to be assessed on a cat-by-cat basis. Consult your local coyote-tracking agency if you are unsure.)

ENGINEERING 101
Moments

"Give me a lever long enough and a fulcrum
on which to place it, and I shall move the world,"
said Archimedes, pretty confident that he would
never have to prove this claim.

The hater from the introduction does bring up a valid issue. In order for the drawbridge to work, it will need a way to close itself. To explain how it will do that we will look at the *moments* applied to the open door.

If you are not already familiar with the concept of a moment, that last sentence might be a little misleading. In this context, a *moment* is an engineering term that has absolutely nothing to do with time. Yes, there is some linguistic relation, but for the purposes of engineering it is best to pretend that you have never heard that word before.

A moment is a force multiplied by the distance to a pivot point, so the units are force times distance. For example, if the weight of a child applies a fifty-pound downward force six feet away from the pivot on a seesaw, there will be a *moment* of *three hundred foot-pounds* at the pivot point (*50 lb × 6 ft = 300 ft-lbs*). If on the other side of the pivot the child has a larger friend who is twice as big, but they sit only three feet away from the center, then that would also apply a three hundred foot-pound moment to the pivot point and the seesaw will be balanced.

The farther away from the pivot point a force is applied, the easier it will be to rotate about that point. This relationship is utilized *all the time*. It is easier to tighten a

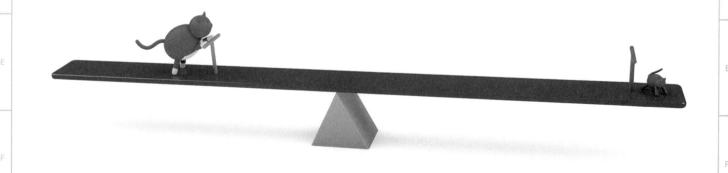

bolt with a wrench that has a longer handle. Construction cranes are balanced with massive weights on the opposite side of the tower. And also some third thing to complete this list, possibly involving spaceships. But most important, this is relevant for our drawbridge door. We want the door to close automatically, but we don't want to put the hinge in the middle—we want it close to the bottom. This will put more mass on one side of the door and that mass will be acting farther away from the pivot point. So in order for the door to close, we need way more weight under the pivot point than above it. The figure below shows the forces acting around the pivot of the door:

Door shown in horizontal (open) position:

We assume that the mass is distributed evenly, so the force from the weight of the door is applied halfway along the length on each side of the pivot.

The force on the main side of the door (F1) is applied 5.625" from the pivot, while the force on the other side (F2) is applied only 1.125" inches from the pivot. This means that for the door to just be *balanced*, F2 would have to be five times greater than F1. Fortunately, steel is about ten times denser than wood, so by stacking steel plates at the bottom we can balance the door without adding too much volume.

Cat Eyes Are Bad at Operating Machinery, Including This Door

Have you ever rewarded your cat for a long day of napping with a treat? Your cat might go crazy trying to get it out of your hand only to knock it onto the floor, where it somehow becomes invisible to him! You can point right at the treat, but unless he sniffs it out he just can't seem to locate it. Why?

The reason is that cats have fairly poor close-up vision. Their large eyes allow for excellent night vision and are worth at least +10 adorableness points, but it takes quite a bit of effort for a cat to focus on close-up objects. They can instead feel things out with their whiskers or paws, but every so often it becomes pretty clear that they are in dire need of some reading glasses. One such situation is with this drawbridge, meaning it may take a while for your cat to use it successfully.

If your cat is treat-motivated, then you may be able to coerce him to learn pretty quickly. Otherwise your cat may have to figure it out on his own over time. If your cat really can't get it, then he probably just does not want whatever is on the other side of the door.

Tools

#	Description	Potential Alternative
T1	Drill and drill bits (capable of drilling through steel)	-
T2	Hacksaw	-
T3	Jigsaw	Drywall handsaw
T4	Wrench	-
T5	Wood glue	-
T6	Circular saw	-
T7	Clamps	Weights
T8	Finishing materials (sandpaper, and paint or stain, if desired)	-

Dimensions are in inches.
Drawings are not to scale.

Materials

#	Description	Size	Quantity
P1	¼" plywood	2' × 2'	1
P2	½" plywood	2' × 4'	1
P3	Steel bar	¼" × 1½" × 36"	2
P4	Threaded coupling	¼-20 thread size, about 1" long	2
P5	Fancy bolts	¼-20 thread size, 2½" long	2
P6	Fancy bolts	¼-20 thread size, 3" long	4
P7	Acorn nut	¼-20 thread size	6
P8	Flat L-brackets	About 1½" along each side	4
P9	Corner braces	About ½" along each side (two fasteners needed per side)	2
P10	Washers	To fit with holes in corner braces	25
P11	Fancy washers	For ¼" fasteners	25
P12	Eye bolt	¼-20 thread size, 2" long	2
P13	Nuts	¼-20 thread size	8
P14	Small wood screws	⅝" or ¾" long	20
P15	Partially threaded bolt	¼-20 thread size, 2" long	2
P16	Badge retractors (see note)	Not heavy-duty, just regular light ones	2

Note: Badge retractors, an optional feature, are those little retractable strings people use to attach ID cards or keys to their belts or personal items. They should be available in a hardware store near the keys.

Phase 1: The Door

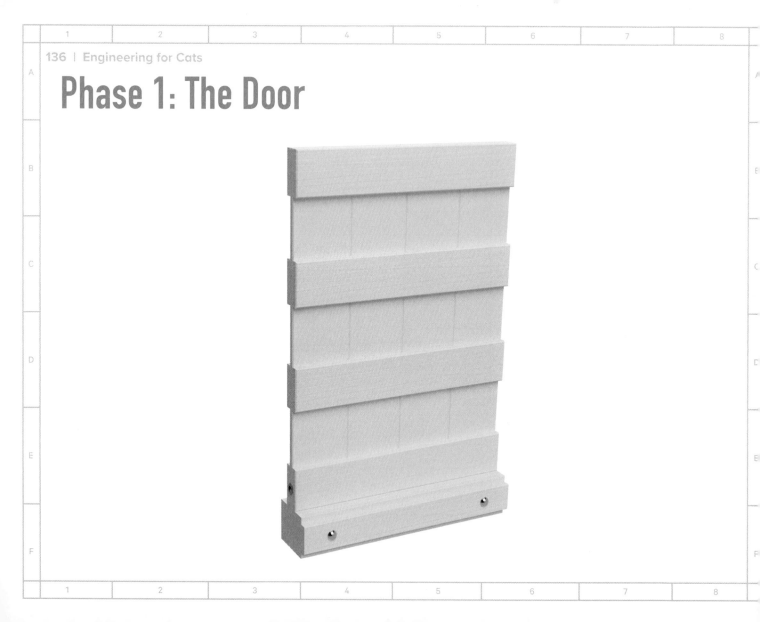

Step 1 Measure, mark, and cut the wood pieces for the door on the ¼" plywood.

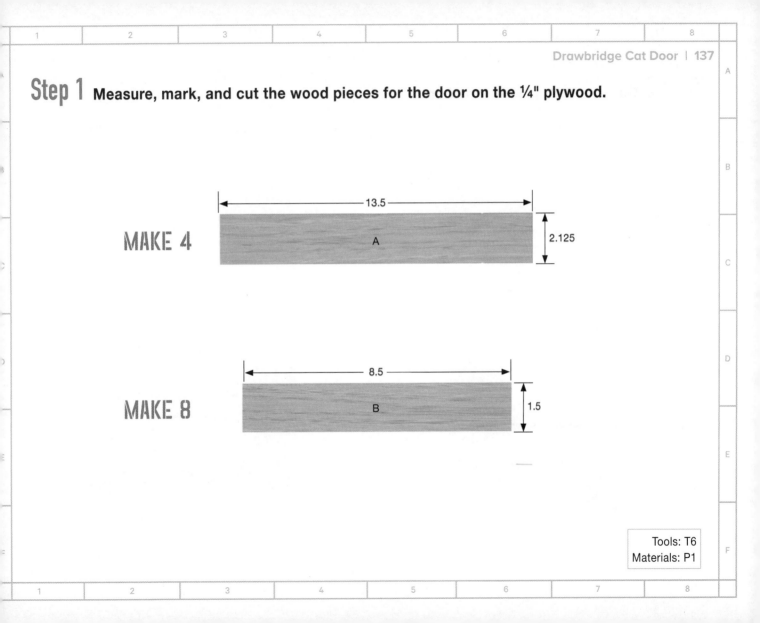

MAKE 4

13.5

A

2.125

MAKE 8

8.5

B

1.5

Tools: T6
Materials: P1

Step 2 Glue the door pieces together.

You may have to glue one side on, let it dry, and then glue the other side. Be sure to apply pressure with clamps or weights to any glued joint as it dries.

Step 3 Add the coupling.

Once the glue dries, drill a hole on each side for the coupling to fit into.

The hole should be halfway through the bottom horizontal piece and as centered as possible. The closer to center it is, the better the door will balance.

Tools: T1, T8
Materials: P4

Step 4 Cut the steel plates.

Cut six pieces of steel bar 8½" inches long. Match drill (see below) ¼" diameter holes as shown to mount the plates to the door.

How to match drill: Drill one steel bar to the dimensions shown below. Then use that piece as a guide to drill the other pieces so that they line up.

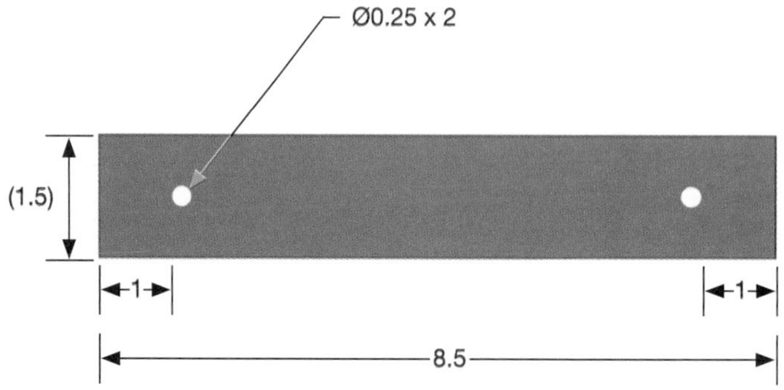

Ø0.25 x 2

(1.5)

|←1→| |←1→|

8.5

Tools: T1, T2
Materials: P3

Step 5 Balance the door.

Match drill a ¼" hole from a steel plate through the bottom of the door. Then add plates to the door until it is balanced where the couplings are mounted (or slightly heavier on the bottom side).

You can check to see that the door is balanced by threading bolts into the couplings, picking the door up from the bolts, and feeling which side is heavier.

Tools: T1, T4
Materials: P5, P7, P11

Phase 2: The Frame

The frame will be the interface between the cat door and the human door. The trim will cover up the cut made in the door so that it doesn't have to be made perfectly.

This phase involves several pieces of wood coming together at different angles. The more precise the cuts are, the easier it will all come together. The assembly does not really require all the joints to have flush contact, so don't worry about small gaps in the frame. If you want to clean up the look you can fill in these gaps with wood putty after assembly.

Step 6 Cut out the pieces for the frame.

Take your ½" piece of plywood and cut out the twelve pieces needed for the frame.

If it is too difficult or you do not have the proper tools to make the angled cuts, you can modify the design to have straight butt joints (shown below as simple joints). This might not look as pretty, but it can be much easier to cut and assemble.

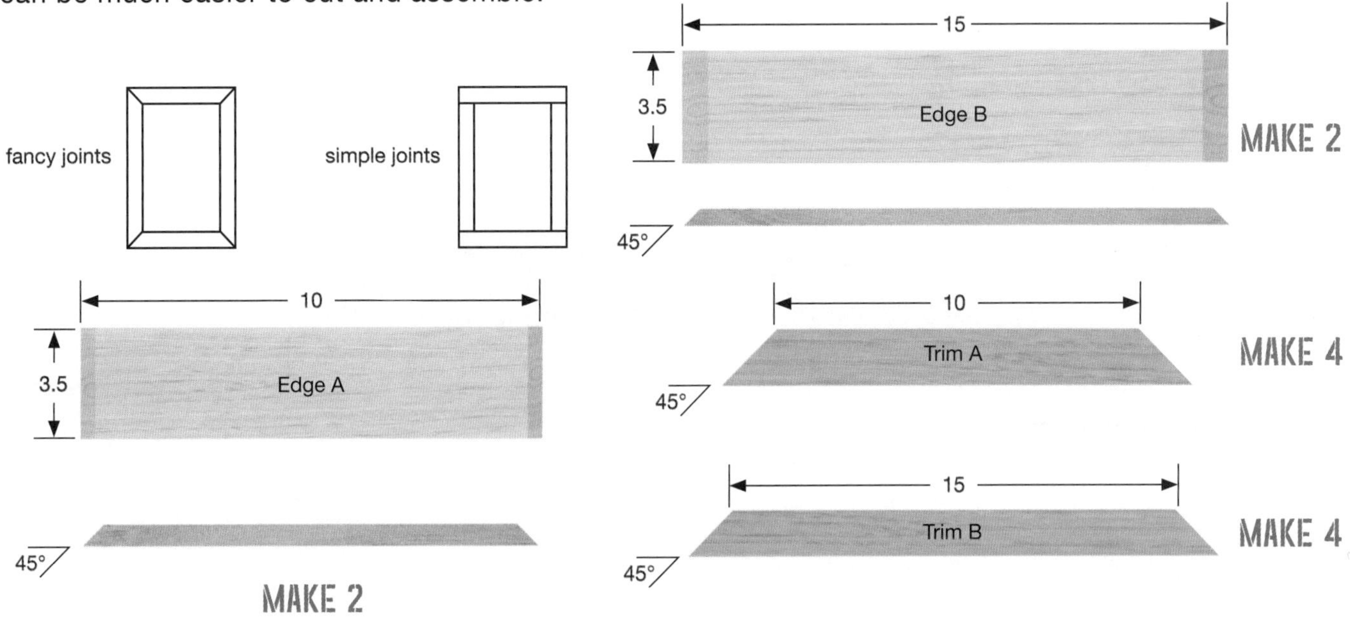

fancy joints

simple joints

15

3.5

Edge B

MAKE 2

45°

10

3.5

Edge A

45°

MAKE 2

10

Trim A

45°

MAKE 4

15

Trim B

45°

MAKE 4

Tools: T6
Materials: P2

Step 7 Drill out the top and bottom trim.

Match drill (see page 140) two holes with diameter 0.35 as shown, through the top and bottom pieces of trim. These will be used to bolt the two halves together when it is time to attach it to the door.

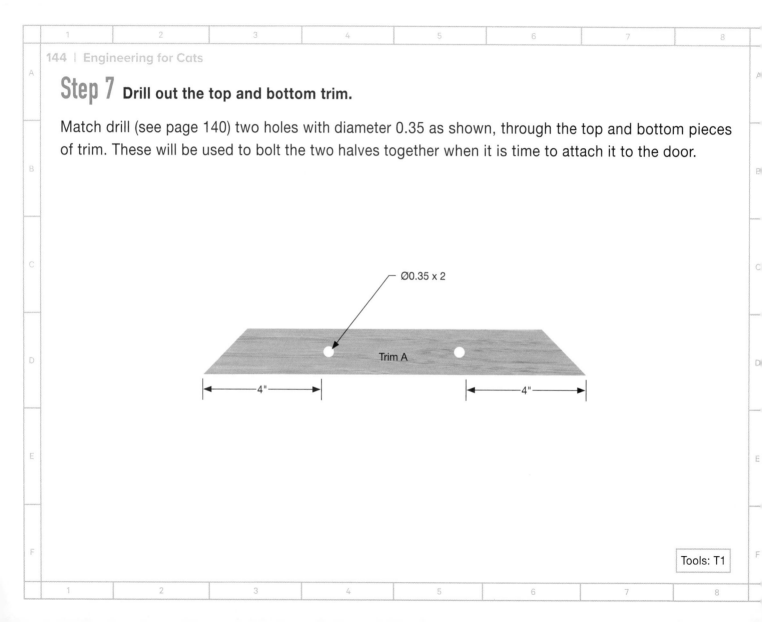

Ø0.35 x 2

Trim A

4" 4"

Tools: T1

Step 8 Assemble the passageway.

Using wood glue and, if necessary, finishing nails, assemble the edges that will create the passageway through the door.

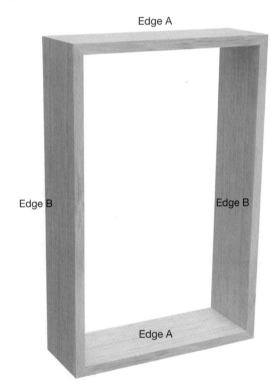

Edge A

Edge B

Edge B

Edge A

Tools: T5, T7

Step 9 Assemble the fixed trim.

The frame has the same style of trim on both sides of the door. This side will be screwed into the passageway that was assembled in the previous step. The other side will be free to slide back and forth over the passageway. This free side will eventually bolt to the fixed side, clamping the human door in between.

The fixed trim can be assembled by screwing two small screws through the passageway and into each edge of the trim.

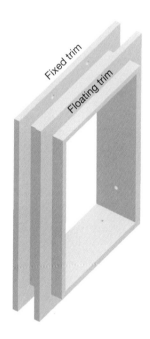

Fixed trim

Floating trim

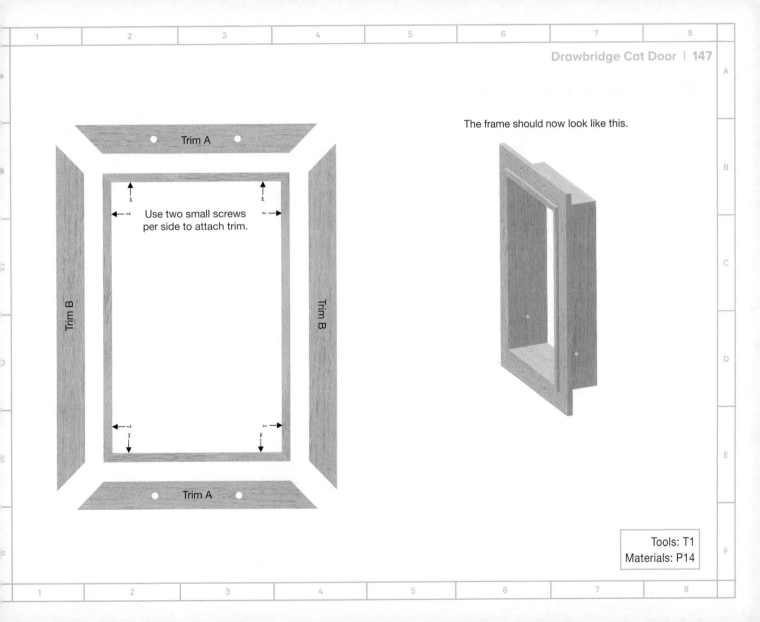

Trim A

Trim B

Trim B

Use two small screws
per side to attach trim.

Trim A

The frame should now look like this.

Tools: T1
Materials: P14

Step 10 Assemble the floating trim.

This trim will not be attached to the rest of the frame yet. Use the flat L-brackets and some wood glue to assemble. Before the glue dries, make sure to check that it fits the passageway.

Trim A

Trim B

Trim B

Trim A

Tools: T1, T5, T7
Materials: P8

Step 11 Drill holes in the passageway to mount the door.

Use the preconstructed door as a guide to locate two holes in the passageway.

These should be clearance holes. The bolts attached to the door will not actually rest on them but should be able to pass through with clearance on all sides. A ½" drill bit should be sufficient.

Ø0.50 x 2

Tools: T1

Step 12 Install the suspension assemblies.

Each assembly will support the bolts that connect to the cat door. For the drawbridge to swing cleanly through the passageway without interference, it will need to be centered precisely. This assembly is designed to allow for adjustment, so the door can still be centered even if the measurements aren't perfect.

To build the assembly:

1. Modify the corner brace by trimming one of the legs, leaving just one of the holes. This will allow it to fit inside the door.

2. Screw the corner brace into the fixed trim side, with a stack of about six washers between the bracket and trim.

3. Secure the eye bolt onto the bottom of the corner brace with nuts on either side.

To adjust the assembly side to side, you can add or subtract washers. To adjust the assembly up and down, you can adjust the nuts securing the eye bolt.

Install this assembly on both sides of the frame.

Tools: T1, T2, T4
Materials: P9, P10, P12, P13, P14

Step 13 Bolt the drawbridge to the door.

First add two washers on the inside and outside of the frame as shown. Then screw the bolts into the couplings mounted on the door.

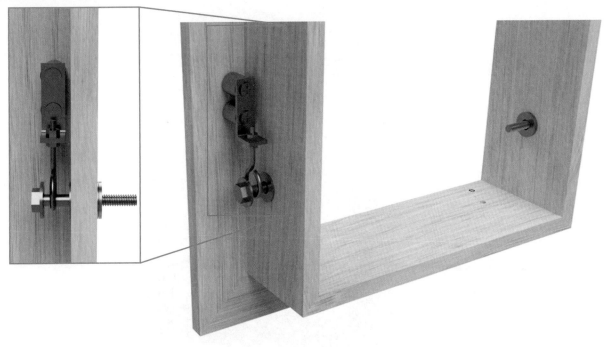

Tools: T4
Materials: P11, P15

Step 14 Adjust the suspension assemblies until the door can swing all the way through the passageway without making contact on any side.

Tools: T1, T4

Step 15
Take off the door and paint the frame and the door separately. Do not paint the bolts or suspension assembly.

Tools: T1, T4, T8

Step 16 Add the badge retractors, if desired.

To help the door rotate and bring it back to vertical every time, two badge retractors can be attached to the bottom. To attach these parts:

1. Drill two small holes in the bottom piece of the passageway, close to where the bolts are keeping the steel plates in place.

2. Take off one side of the steel plates but leave the bolt in place.

3. Thread the string from the badge retractor through the holes and tie the end to the bolts. You will have to cut the end of the string, so be careful not to let it snap back into the body of the badge retractor. To prevent this, you can extend the string, and then tie a nail near the body to prevent it from reeling in the string while you work.

4. Reattach the steel plates, sandwiching the end of the string.

The badge retractors can just hang underneath the bottom edge of the passageway. They will be concealed in the door when the final assembly is attached.

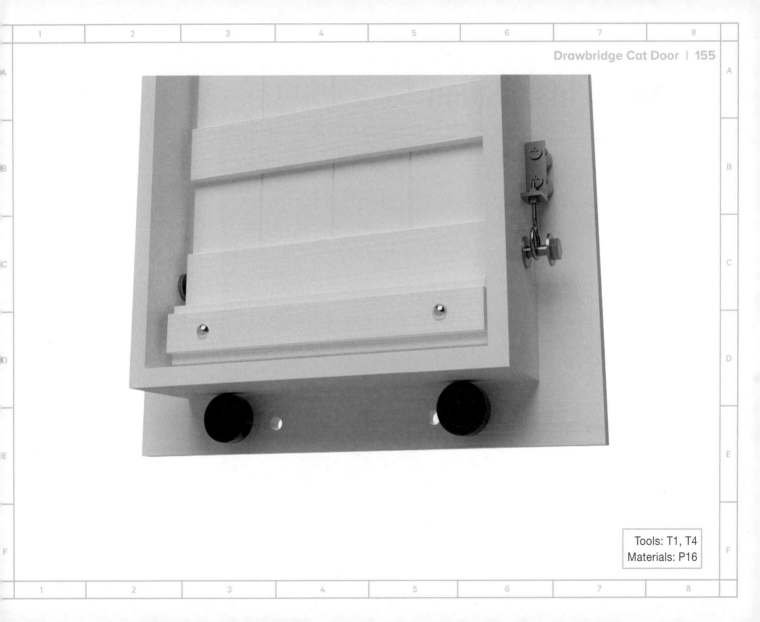

Tools: T1, T4
Materials: P16

Phase 3: Installation

Step 17 Cut the hole in the door.

Take the human door off its hinges. Use the passageway as a guide to mark out the cut, leaving some space for the suspension assemblies and badge retractors. It is best to make this cut slightly large, because the door will be secured on bolts and it's okay if the fit is loose.

Cutting into a nice, functional door can be scary. But don't worry. The trim of the drawbridge will cover up almost any jagged edges, so the cut does not have to be super clean. And if you do screw up so badly that the door is unusable, hardware stores usually carry spares.

Bonus: If you have a hollow door, the badge retractors and suspension assemblies will fit in much more easily.

Tools: T3

Step 18 Drill the bolt holes in the door.

Slide the passageway into the door and match drill (see page 140) through the trim.

Match drill all 4 holes
through the door.

Tools: T1

Step 19 Bolt the doors together using the fancy washers and acorn nut.

You will have to trim the bolt with a hacksaw in order for the acorn nut to pull tight against the door. When it's secure, return the human door to its rightful place.

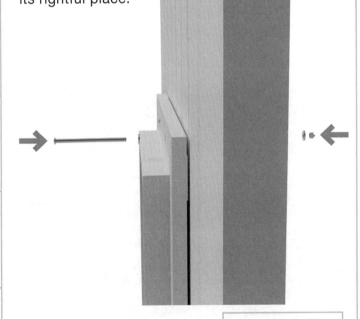

Tools: T2, T4
Materials: P6, P7, P11

Complete!

LITTER BOX CABINET

Problem

Despite the little maintenance that cats require, they do still need access to a litter box. Litter boxes are often unattractive waste disposal mechanisms that humans don't want to deal with.

The Cats vs. Dogs Debate

One of the more practical selling points of cats is their cleanliness. You don't usually have to bathe them, they rarely leave dishes in the sink, and they come with a genetic predisposition to use a litter box. These are pretty special qualities to have in a pet. In fact, these are usually the main arguments when comparing them to their canine counterparts. Dogs provide a level of loyalty well beyond what you would ever expect from another human; they can be trained to do all sorts of jobs from hunting to aiding

the disabled, and any size and style imaginable can be procured. When it comes to which creature can do the most to earn its keep, the score is a little bit lopsided.

This almost complete deficiency of useful skills apparently does not matter for many pet owners, as cats are currently the most popular house pet in the United States, with a population approaching 100 million. There are a few other distinct advantages that cats hold. They do better in city apartments, they can be left alone for longer periods of time, and they can respect that maybe you're just trying to get rid of a tennis ball and you don't want it back. The heavy hitter for many people, however, is the low maintenance.

Though cats have very few practical uses, their caretakers are considerably less burdened by their presence.

Despite having the best option for waste disposal among house pets by a wide margin (birds were a close second until the newspaper industry went under), some people still underestimate the effort involved in maintaining a litter box. This is understandable. The nice thing about cats is that you don't have to provide bathroom services for them several times a day like dogs. You really only have to take action a few times a week, but this can make it easy to put off. Two days can quickly turn into five, and in that time span what would have been a thirty-second chore can explode into a full-blown hazmat situation.

Solution

Build a litter box that's easier to maintain on a regular basis, and more enjoyable for your cat to use. The considerations for the cabinet design can be useful for any litter box setup.

CONSISTENCY

Scooping the litter box on a regular and frequent interval makes things much easier than scooping only when the box becomes a noticeable problem. There will be less to clean each time, and it will be more enjoyable for your cat to use. If waste builds up you will have to replace the litter more often, which is more expensive, and because it is more time-consuming you will want to do it even less often, making it a worse chore than it has to be.

This box encourages consistency in several ways, but most directly through the tracking system. On the front of the cabinet are seven hooks, each one labeled with a day of the week. After you scoop the litter, you put the tag on the current day. This will provide a visual indicator of how long it has been since the last time it was scooped. Ideally this would be every one or two days. Because it will be easy to forget about the litter until the situation becomes dire, this can help provide an early warning and motivate you to clean it while it is still manageable. If you cannot tell if it has been zero or seven days since you scooped the box, you should probably take your cat to the vet or yourself to the hospital.

LOCATION

By placing the box in a convenient location you can also make it easier to scoop consistently. It is not wrong to want to place the box in a hidden location, and you should be able to tuck away this cabinet as well, if desired. But if it's too well hidden, it will be easy to forget about. Using a cabinet like this can make it easier to hide the litter box without making it inaccessible to clean or uncomfortable for your cat to use.

If you have multiple cats, it can really make their lives easier if you also get multiple litter boxes. You may have used up your only discreet spot available on the first box, so the cabinet can provide an inconspicuous option that would have otherwise been too exposed.

Cost:
High

Difficulty:
Low/Medium

Estimated Build Time:
6 to 8 hours

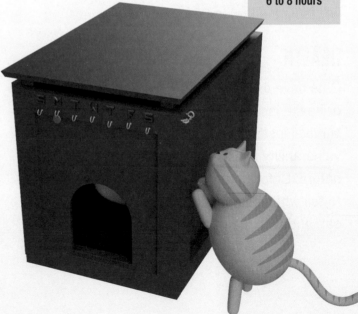

CONVENIENCE

The amount of time it takes to do a chore is usually directly related to how often it gets done. By building this cabinet you can also conveniently place all the necessary tools and materials, such as extra litter, plastic bags, and a scooper, right there. This may save only seconds per cleaning, but it will add up. The convenience will facilitate the consistency.

HEALTH

Cats have a natural tendency to conceal pain or health issues. This is most likely part of their survival instincts, because announcing your vulnerability is usually not rewarded kindly in nature. Unfortunately, this makes it a little difficult to tell when something may be afflicting your cat, and with many health issues, the earlier it is diagnosed the easier it is to fix. The litter box is one reliable indicator of when something may be wrong. Changes in how cats use it or in what is coming out of them may be the first sign of distress, even if they are otherwise acting fairly normal. If you do not clean their litter box enough, or if you use one of those automatic scooping machines (sometimes referred to as a "human child"), then you are unlikely to recognize when something is wrong.

The litter box may be the least attractive part of cat ownership, but compared to the alternatives it is pretty convenient. If set up and maintained properly it really doesn't have to be as big of a chore as some people make it out to be, and your cat will appreciate it more than anything (although she certainly won't show it).

ENGINEERING 101
Ockham's Razor

Ockham's razor is a problem-solving rule dating back to the nineteenth century, attributed to thirteenth-century philosopher William of Ockham. It is often referenced as an important design principle in engineering. It basically states that if you have multiple proposed solutions to a problem, the one that relies on the fewest amount of assumptions should be selected. A modern application of the principle is to say that the simplest solution to a problem will likely be the most effective.

This, of course, is utter garbage. It is really only a good design philosophy if you want to end up with a boring design, and it is probably why 99 percent of the products that could have lasers on them don't. I want to know how this Ockham guy gets credit for such a basic and widely applicable philosophy, anyway. Did he dedicate his life to scholarly pursuits, developing modern thought of the time by challenging other commonly accepted philosophies and proposing his own explanations? Or did he plagiarize it and receive credit because he was a loudmouth who wanted to take credit for depriving modern products of extra features? According to his own principle, it seems that this medieval friar was a loudmouth with a grudge against lasers. Undeniable.

I hate to admit that one important application of Ockham's razor is when the design in question is meant to influence human behavior. In the case of the litter box, the goal is to prevent your cat's digestive finish line from becoming a burden in your home. No matter how fancy your solution to this problem is, eventually it will require human involvement. There are many ways in which this could be approached and many features that could be added, but ultimately most of these options would just add complexity. The objective with this cabinet is to create a regular habit out of cleaning the litter box. If you

add complexity to a task that you are not looking forward to already, you probably will be making it less likely that the task gets done (with just a few laser-related exceptions). Fortunately, scooping litter is not complex, and if done regularly it can become a habit and will no longer be on that long list of things that you dread doing every day.

Tools

#	Description	Potential Alternative
T1	Jigsaw	Circular saw, handsaw
T2	Drill and drill bits	-
T3	Finishing materials (sandpaper, paint, stain . . .)	-

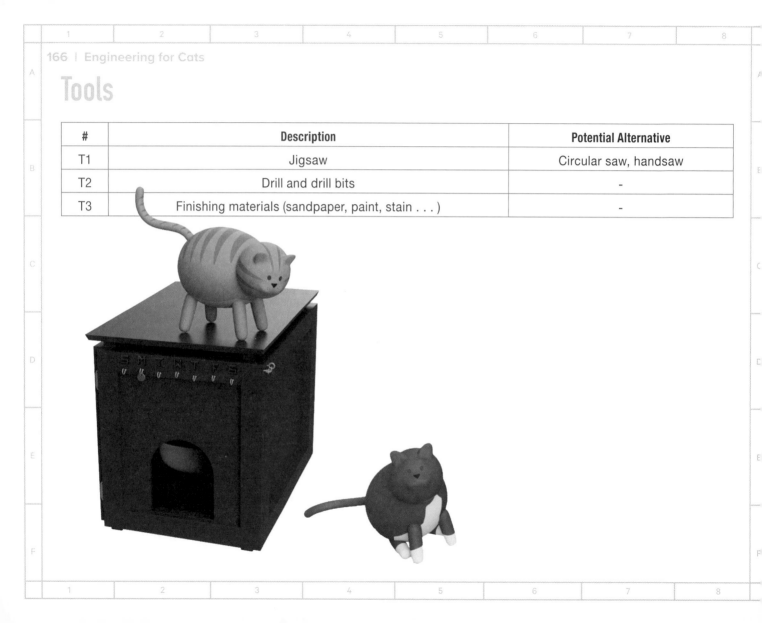

Materials

#	Description	Size	Quantity
P1	2 × 2 stud	8' piece	2
P2	½" plywood	4' × 8' piece	1
P3	¾" plywood	4' × 8' piece	1
P4	1 × 3 plank	8' piece	4
P5	Wood glue	1 bottle, but probably not the whole thing	1
P6	Medium wood screws	Between 1¼ and 1½" long	1 pack (≈50)
P7	Corner braces	1" legs	8
P8	Gate hook	Hook and eye gate hook for keeping cabinet closed, 3"	1
P9	Hinges	Small cabinet hinges for flush mounting, approx. ⅜"	2
P10	Hooks	Self-threaded brass hooks, ¾"	8
P11	Letters	Adhesive letters for labeling days	1 pack
P12	Caulk and caulk gun	For filling in gaps between pieces (optional)	1 tube

Dimensions are in inches.
Drawings are not to scale.

Step 1 Measure, mark, and cut four pieces of 2 × 2 for the legs.

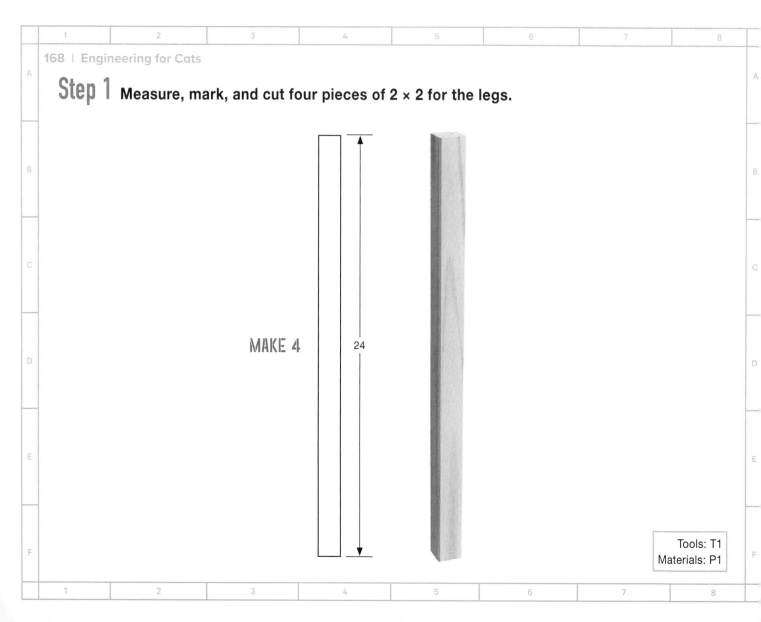

MAKE 4

24

Tools: T1
Materials: P1

Step 2 Measure, mark, and cut the sides out of ½" plywood.

These dimensions are sufficient for a large cat to use and to store litter and other supplies inside. It can be made smaller to fit into tighter spots, or to take up less room if your cats aren't overweight panthers like mine. Adjust the dimensions based on where you plan to place the cabinet.

BACK

MAKE 1

21

22

SIDES

MAKE 2

28

22

FRONT

MAKE 1

20

22

11

8

Tools: T1
Materials: P2

Step 3 Cut the trim from the 1 × 3 to match the sides that you cut in Step 2.

Measure and mark the trim to match the dimension of the sides. The dimensions listed are assuming you cut the sides correctly.

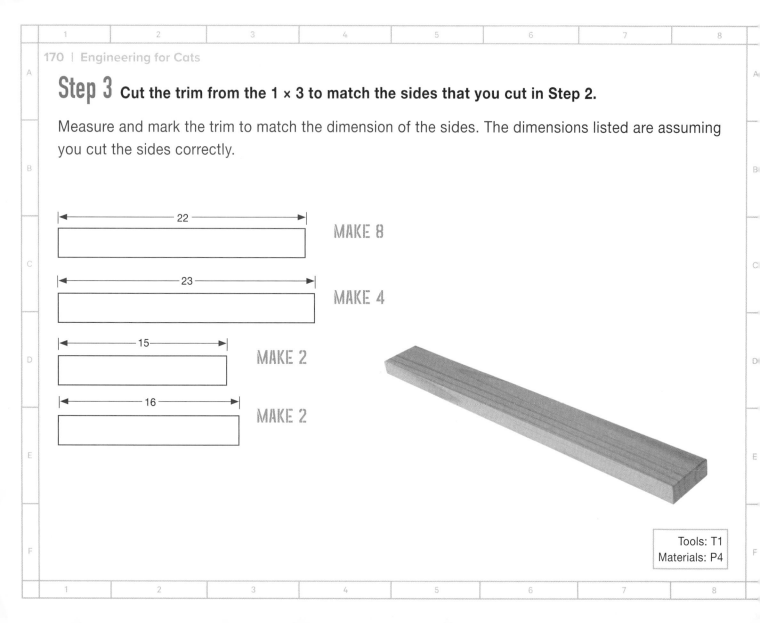

|← 22 →|

MAKE 8

|← 23 →|

MAKE 4

|← 15 →|

MAKE 2

|← 16 →|

MAKE 2

Tools: T1
Materials: P4

Step 4 Glue the trim to the sides.

This will significantly increase the fanciness of the cabinet.

Materials: P5

Step 5 **Assemble the first three sides.**

Use the medium wood screws to attach the legs to the sides first, lining them up with the edges.

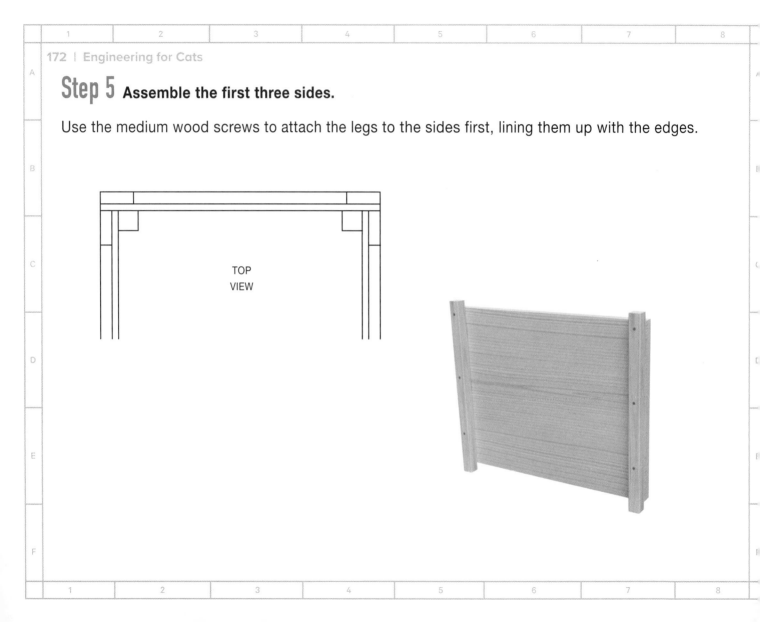

TOP
VIEW

Then attach the back so that it lines up with the edge of each side as shown.

Tools: T2
Materials: P6

Step 6 Measure, mark, and cut the bottom piece from ¾" plywood and install it with corner braces.

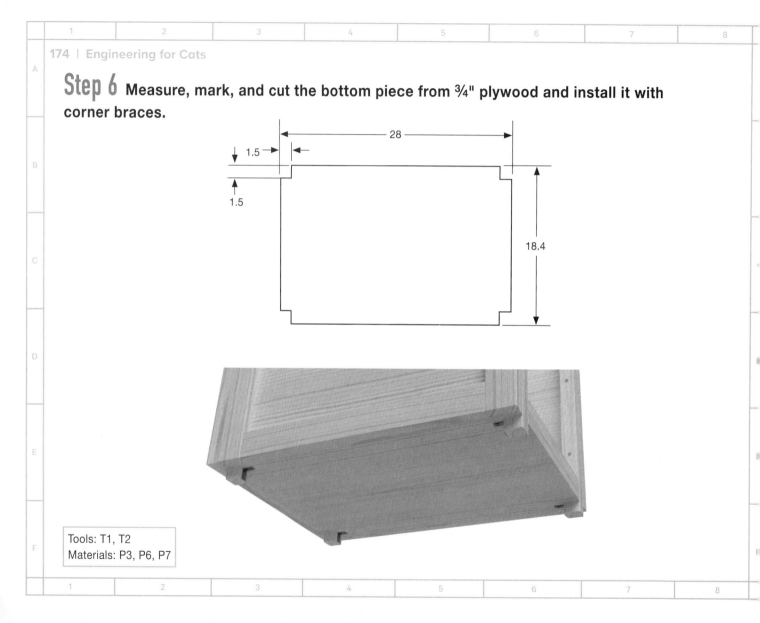

Tools: T1, T2
Materials: P3, P6, P7

Step 7 Measure, mark, and cut one more piece of 2 × 2 and install as shown.

This piece will help prevent litter from spilling out the door. If there are any gaps between the sides and bottom piece (which is expected), use caulk to fill it in. This will also help prevent litter from spilling out.

15.4

Tools: T1, T2
Materials: P1, P6, P12

Step 8 Measure, mark, and cut the top piece from ¾" plywood.

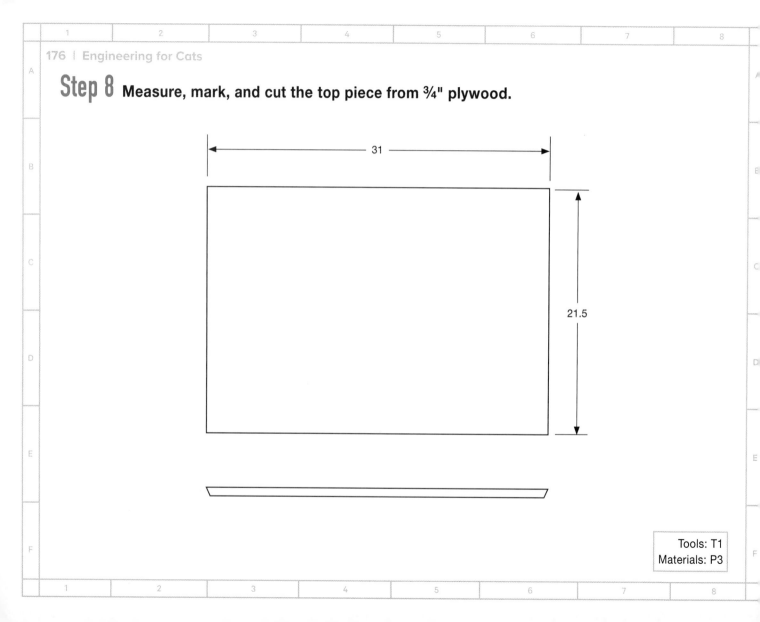

31

21.5

Tools: T1
Materials: P3

Step 9 Paint the assembled sides, front, and top piece, and let dry.

Tools: T3

Step 10 Attach the top using the corner braces.

Tools: T2
Materials: P6, P7

Step 11 Install the door with the hinges and latch.

Tools: T2
Materials: P8, P9

Step 12 Install the hooks and letters for daily scoop tracking.

Tools: T2
Materials: P10, P11

Complete!

CAT WHEEL

Problem

A hyperactive cat may be full of excess energy that she is taking out on you, the furniture, or your friends and family . . .

On the other hand, some cats may have the opposite problem—lethargy. And while exercise is beneficial for any cat's healthy lifestyle, many cat owners mysteriously decline to take their feline companions outside for a game of fetch.

An important lesson in self-confidence.

Solution

Other projects in this book such as the shelves and slot machine can help get a restless and destructive cat in control, but the wheel will really get her moving. Based on a product commonly made for hamsters, this wheel is a great way to exercise an overactive cat, or help get an especially "fluffy" cat to a healthier weight.

There are commercial alternatives available for purchase, but they can be quite expensive (up to $1,000), and the options are limited. The materials for this wheel are estimated at just under $200, so although not exactly cheap, financially this is a much less risky option if you are unsure how your cat will take to the wheel. If, in the worst-case scenario, your cat completely rejects the wheel, you can donate it to your local animal shelter. There will be plenty of animals looking for an opportunity to stretch their legs there.

Cost:
High

Difficulty:
High

Estimated Build Time:
10 to 20 hours

ENTERTAINMENT DISGUISED AS EXERCISE

There are at least several differences between hamsters and cats. I'm not claiming to be able to tell them apart every time—the majority of animal DNA is just cut and paste anyway, so that seems unrealistic—but my understanding is that there are some differences. Fortunately, this project focuses on something that they have in common: the ability to expend energy running on a wheel.

As with anything you get for your cat, this wheel may end up becoming a big part of his life that he uses all the time, or it may be something he aggressively ignores. So before you dive into this project, you might want to evaluate what your cat's motivation might be. The type of cat that will get the most use out of this wheel is a hyperactive cat looking for an energy outlet. These are usually young cats or those of an energetic breed such as a Bengal. These types of cat may take to the wheel with little encouragement and may eventually use it of their own volition, just like a hamster. The second type of cat that could really benefit from this wheel is one that is overweight and looking to shed a few pounds. This is a good cat to build the wheel for, but it will be more difficult to get him to use it.

Body image for a cat is actually not a strong motivator, and in my experience no amount of fat shaming will influence their behavior (also similar to hamsters). In these cases you can use toys and treats to encourage your cat to run on the wheel, but he may never just go for a run on it by himself.

ENGINEERING 101

Circles

Many of the factory-made cat wheels available for purchase look slightly different from this one. They have the entire wheel resting on rollers, while the one proposed here is more of a Ferris wheel–type mount. This roller design is great because it is open on either end, making it a little easier to use, and it requires no support in the middle so the base structure is much simpler.

So why does this proposed wheel not take advantage of this design? The reason is that the roller design requires circles, big ones, and big circles are almost impossible for humans to make. If these circles have even small imperfections, they would wobble as they rolled, making it very unsettling for your cat. Making circles that roll properly usually requires expensive machinery, and making them large enough for this project would be very pricey. This is true in industries outside of the cat/hamster-wheel market as well. Think about all of the factory-made circles you encounter every day: frying pans, bicycle wheels, steamboat paddles. The costs shoot up as the circles get bigger.

Circles have a unique ability to blow up in cost as they get bigger for good reason. Consider a length of wood that you want to cut accurately within $\frac{1}{16}$ of an inch (0.0625 inches). This matter is simple if the final length is around 6" total: the two edges are close together and a variety of tools can be used to verify the length. But if you are cutting it to be 60" long, it would take much more effort to achieve this precision. This effect is compounded many times over for circles. Determining how circular something is requires much more effort than checking one length measurement. You would have to check the dimensions in many spots to make sure the circle did not come up elliptical, and there are no well-defined edges to reference.

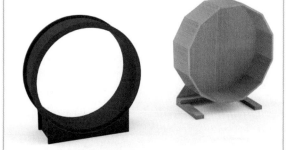

Two types of cat wheel design

With this circular dilemma, the design of this project is actually much simpler because it avoids having these large circles at all. It does require enough precision so that the wheel is centered and balanced, but small imperfections will have minimal impact on the function.

Tools

#	Description	Potential Alternative
T1	Circular saw	Table saw and jigsaw
T2	Drill and drill bits	-
T3	Scissors	Box cutter
T4	Staple gun and staples	Hammer and tacks
T5	Wrench	-

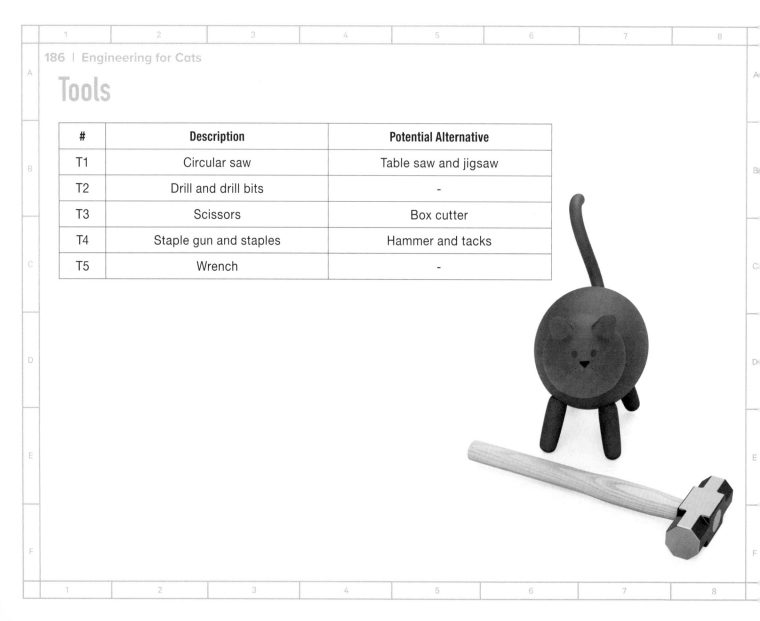

Materials

#	Description	Size	Quantity
P1	⅜" plywood	4' × 4'	1
P2	⅝" plywood	4' × 8'	1
P3	2 × 4 beam	8' long	2
P4	Wood screws	1¼" long	100
P5	Wood screws	2½" long	25
P6	L-brackets	About 2½" arm length	12 to 24
P7	1 × 12 wood board	8' long	2
P8	Carpet (see Note)	8' × 6'	1
P9	Carpet tape	2" Wide	1 roll
P10	Bolts	¼-20 thread, 1½" grip length	4
P11	Flat washers	To fit ¼-20 bolt	4
P12	Nuts	¼-20	4
P13	Large swivel caster	See Step 8 for details. Caster width will have to match width of 2 × 4 (1½") so the wheel can be mounted to the base.	1
P14	Small fixed caster	Approximately 2" wheel. See Step 13 for details.	2
P15	Sisal rope	⅜" diameter	300 feet

Dimensions are in inches. Drawings are not to scale.

Note: The ability to withstand clawing varies wildly between types of carpet. Make sure to get a compact, tightly woven type. Shag or something with large loops will start to look haggard very quickly.

Phase 1: Constructing the Wheel

Step 1 Measure, mark, and cut out twelve panels to form the wheel using the 1 × 12 board.

If you do not have the tools to cut the 15° angles on the ends, you can cut them flat. Overall the wheel will look and function just the same in the end. If the ends are cut flat, use two L-brackets per panel in Step 3 instead of just one.

15°

0.75

10.25

MAKE 12

Tools: T1
Materials: P7

Step 2 Measure, mark, and cut out the backing panel from the ⅝" plywood.

First arrange the 12 panels cut in Step 1 into a twelve-sided polygon (dodecagon), making sure that it is even and not skewed in any dimension. To check that the polygon is even, measure the distance between each pair of boards that face each other. All six of these distances should be equal.

If the 15° angles are cut in Step 1, it might be helpful to clamp the pieces together with a ratchet strap or rope.

Trace this shape onto the plywood, and offset the cut so that the backing is ¾" inches larger around the edge.

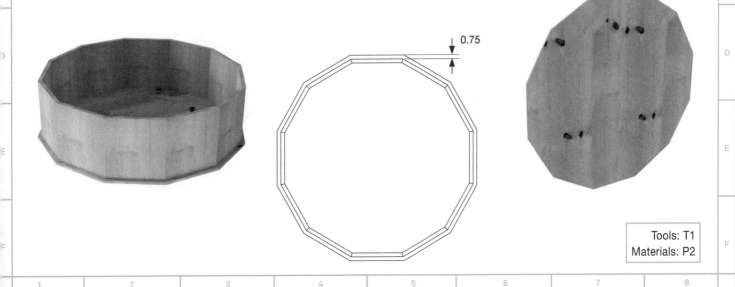

0.75

Tools: T1
Materials: P2

Step 3 Attach the panels to the backing with L-brackets.

Tools: T2
Materials: P6

Step 4 Measure, mark, and cut out the front trim piece from ⅜" plywood.

This piece should match the outer profile of the backing piece, and should have a width of 2.25 inches.

2.25

Tools: T1
Materials: P1

Step 5 Lay down carpet on the backing.

Use a healthy amount of carpet tape to secure it and then staple around the perimeter.

Tools: T3, T4
Materials: P8, P9

Step 6 Lay down carpet on the track.

Cut so that it can fold over the front edge, which will be covered by the trim. Use a healthy amount of carpet tape. Add staples to the inside corner and also on the outer edge where it is folded over.

It may take more than one strip of carpet to make the full circle.

Add staples around the inner corner and outer edges.

Tools: T3, T4
Materials: P8, P9

Step 7 Attach the trim. Sand and paint prior to attaching if desired.

Pre-drill six holes on every other panel for the trim. Use small wood screws to attach it.

Tools: T2
Materials: P4

Step 8 Attach the caster plate.

The wheel will turn using a large swivel caster. The caster wheel will be removed—this is where the base will attach later. The turning of the wheel will actually be executed via the bearing in the caster that's meant for swiveling.

Take great care in making sure the caster is centered on the wheel. If it is off center, then the wheel will be unbalanced, making it very difficult to turn, and your cat certainly does not need another excuse not to exercise.

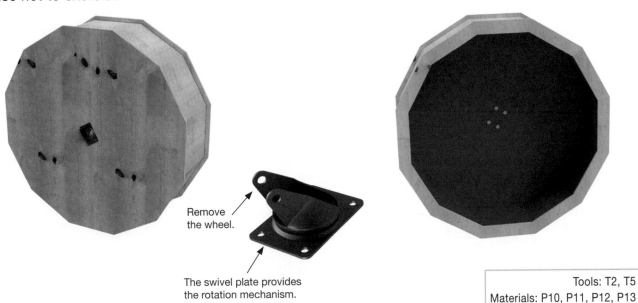

Remove the wheel.

The swivel plate provides the rotation mechanism.

Tools: T2, T5
Materials: P10, P11, P12, P13

Phase 2: Constructing the Base

Step 9 Measure, mark, and cut the appropriate lengths from the 2 × 4.

MAKE 1
|←———21———→|

MAKE 1
|←——————————40——————————→|

MAKE 2
|←———20———→|

MAKE 2
|←——19.25——→|

MAKE 2
|←—7—→|

Tools: T1
Materials: P3

Step 10 Assemble the 2 × 4 frame.

Use the 2½" wood screws where shown.

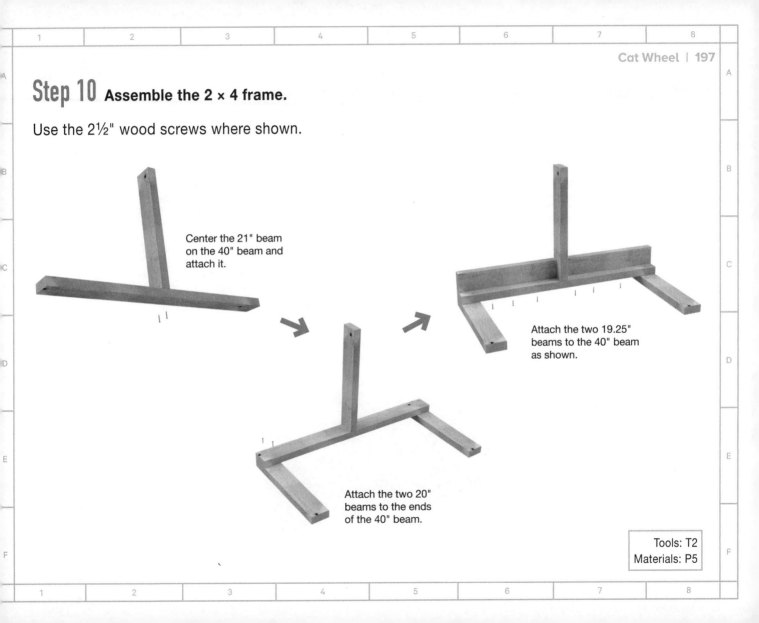

Center the 21" beam on the 40" beam and attach it.

Attach the two 19.25" beams to the 40" beam as shown.

Attach the two 20" beams to the ends of the 40" beam.

Tools: T2
Materials: P5

Step 11 Measure and mark the backing on the ⅝" plywood and cut it out.

Add the two 7" pieces as shown, which will support the small casters in Step 13.

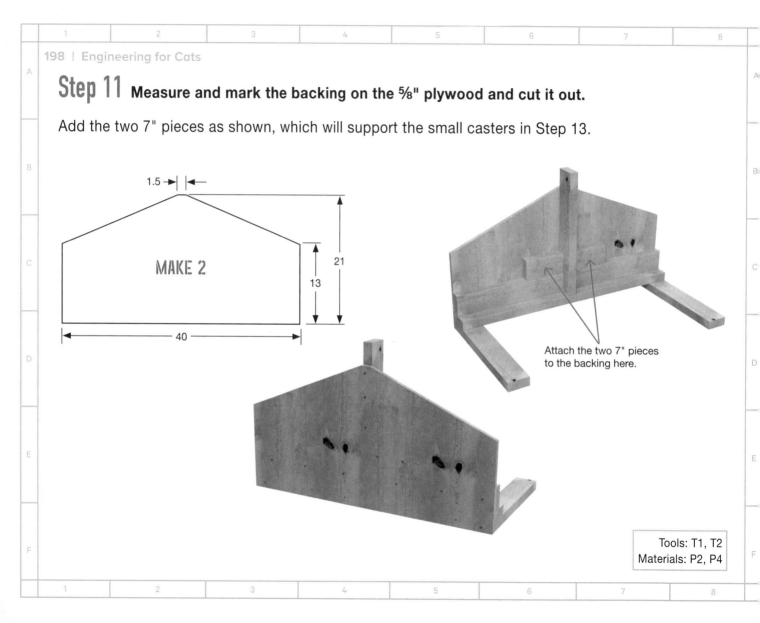

MAKE 2

1.5

21

13

40

Attach the two 7" pieces
to the backing here.

Tools: T1, T2
Materials: P2, P4

Step 12 Measure, mark, and cut out the diagonal supports.

Attach them to the frame with wood screws. Use the shorter wood screws through the plywood on the back, and the longer wood screws through the 2 x 4 beams on the bottom.

45°

16

MAKE 2

Tools: T1, T2
Materials: P3, P4, P5

Step 13 Paint the base, if desired, and then attach the small casters.

These will likely have to be shimmed—supported with thin strips of scrap wood inserted in any gap or opening between the caster and the plywood base—in order to attach them properly. The casters should support the wheel so that it rests vertically and they should be angled so that they roll smoothly on the wheel. No gap will exist if the shims are not used, but the wheel will not sit vertically.

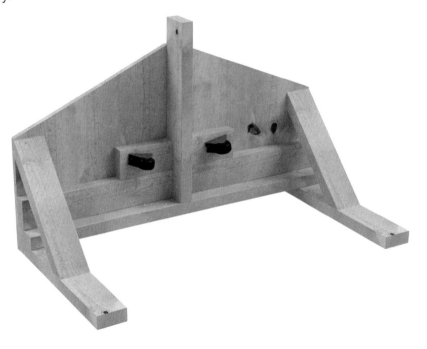

Tools: T2
Materials: P4, P14

Step 14 Attach the wheel to the base.

Drill a hole through the top of the base and mount the wheel using the existing hardware with the caster.

Tools: T2, T5

Step 15 Wrap the wheel in sisal rope—this will actually provide some structural support, as well as another good place for scratching.

Keep constant tension on the rope as you wrap it around the wheel. Fasten with a few staples on the end of each rope, and with a few in the middle to keep everything contained.

Tools: T4
Materials: P15

Complete!

About the Author

Mac Delaney earned his Bachelor's and Master's degrees in engineering from the University of California, San Diego, in 2012. Since then he has been working in the aerospace industry on commercial aircraft and spacecraft structures. He first realized the potential for applying engineering principles to cat-related projects when he discovered that cats are significantly less stressful to interact with than humans. Motivated to learn more, he began to research these peculiar animals, how they became such a common household companion, and what can be done to make that companionship better. In an attempt to contribute to the growing field of cat research Mac takes regular trips to the zoo, where he meows at various animals to see which ones will react. So far none have, but the work is ongoing. Mac is also married to a completely real and not imaginary person. His wife is also an engineer who is supportive of the cat-based projects and tolerant of the zoo-based research. In the future he plans to continue his parallel careers in aerospace and cat engineering, in the hopes that one day these fields will be fully unified. Visit EngineeringforCats.com for more tips on building, photos of the projects, and more.

ACKNOWLEDGMENTS

I would like to acknowledge that this book only exists because of the encouragement and guidance I have received. This has come from too many people to name, but to my entire family, those at UC San Diego, Workman Publishing, and those I have worked with in the aerospace industry— thank you. This book is more a reflection on the support you have given me than anything else.